Automated Broad and Narrow Band Impedance Matching for RF and Microwave Circuits

Amal Banerjee

Automated Broad and Narrow Band Impedance Matching for RF and Microwave Circuits

Springer

Amal Banerjee
Analog Electronics
Kolkata, West Bengal, India

ISBN 978-3-319-99000-2 ISBN 978-3-319-99001-9 (eBook)
https://doi.org/10.1007/978-3-319-99001-9

Library of Congress Control Number: 2018952349

This Springer imprint is published by the registered company Springer Nature Switzerland AG
The registered company address is: Gewerbestrasse 11, 6330 Cham, Switzerland

This book is dedicated to:
My late father Sivadas Banerjee
My mother Meera Banerjee
My sister Anuradha Datta

A dear friend, mentor and guide
Dr. Andreas Gerstlauer
and
Dr. Melvin Oakes and Dr. C Fred Moore
who taught me the physics of RF and microwaves

Preface

Correctly implemented impedance matching guarantees maximum or optimum signal power transfer from a signal source to a load. Clearly, it is essential for the correct operation of any modern electronic signal processing circuit (consumer electronics, industrial process control electronics, medical electronics, and most importantly telecommunication related electronics). Without correctly implemented impedance matching, a significant fraction of the incident signal (on the load) energy and power would be reflected back to the source.

Of the two types of impedance matching, broad and narrow band, the former is more important, as a significant fraction of electronic signal processing circuits operate over a frequency band or range, rather than a single frequency (narrow band). However, design calculations for both broad and narrow band impedance matching sub-circuits are complicated, multi-step, and therefore time-consuming and very error-prone. This is especially true for traditional broad band impedance matching schemes, which are governed by both *conjugate matching* (as in narrow band impedance matching) and the *Bode-Fano inequalities* that impose strict upper limits on the gain and bandwidth of the impedance matching sub-circuit. Similarly, traditional narrow band impedance matching techniques make extensive use of the Smith chart. Using a Smith chart to extract meaningful results requires a high level of proficiency – a result of years of use. All the complexities of impedance matching sub-circuit design calculations arise from the frequency dependency of the reactance ("resistance") of both the capacitor and inductor. To complicate matters, the most load impedances are complex, that is, consist of a parallel or series capacitor or inductor with a resistor. Although some extremely powerful and effective design schemes (Zobel sub-network, active impedance matching) are used to neutralize or nullify the frequency dependency of capacitor and inductor reactances, design calculation complexity is reduced only a little.

Design calculation automation counters these issues. The main chapter in this book (Chap. 4) demonstrates how simple C computer language programs can perform design calculations for each of the broad and narrow band impedance matching schemes examined in Chaps. 2 and 3. The main advantage of automation is that the computed component values (for an impedance matching sub-circuit) are formatted as a text SPICE input netlist. SPICE (Simulation Program with Integrated Circuit Emphasis) is the gold standard in electronic circuit performance evaluation. In this case, the performance characteristics (RMS power transferred from signal source to load, reflection coefficient, insertion loss, and standing wave ratio (SWR)) can be immediately estimated for any new impedance matching sub-circuit, with any available open-source or proprietary SPICE simulator. This allows the designer to explore the design space and fine-tune an impedance matching sub-circuit design. Design accuracy is guaranteed, and design effort and time are cut down significantly. These exciting and innovative techniques are examined in this book.

Kolkata, West Bengal, India Amal Banerjee

Contents

1 Introduction and Problem Statement 1
1.1 Introduction 1
1.2 Why Impedance Matching Is Essential 2
1.3 How to Implement Impedance Matching: Conjugate Matching Condition 4

2 Narrow Band Impedance Matching 7
2.1 Introduction 7
2.2 The Common *L* Impedance Transformer with Complex Load and Source Impedances 7
2.3 Single Transmission Line Segment to Match Complex Load–Source Impedances 9
2.4 The Concept of Reflection Coefficients 10
2.5 The Lambda/12 Transmission Line Impedance Transformer with Complex Load–Source Impedances 11
2.6 Single Series or Shunt Closed/Open Transmission Line Impedance Transformers 13
2.7 Double Series or Shunt Closed/Open Transmission Line Impedance Transformer 14
References 16

3 Broadband Impedance Matching 17
3.1 Introduction 17
3.2 Bode–Fano Conditions on Gain Bandwidth of Broadband Impedance Matching Networks 17
3.3 Converting Complex Series RC or RL or Parallel RC or RL Load into Pure Resistance: The Zobel Network and Parallel to Equivalent Series Impedance Transformation 19
3.4 Active Impedance Matching and Neutralizing Reactances or Susceptances of Complex Impedances or Admittances 21

3.5 Low-Pass Configuration Matching Network Design with Bode–Fano Constraints and Table Lookup 24
3.6 Impedance Matching of Equal Real Source and Load Resistors with Load Having Nonzero Reactance Low-Pass and Band-Pass Matching Network 26
3.7 Norton Series and Shunt Transforms for Matching Unequal Load Source Impedances 27
3.8 Generalized Norton Transforms for Matching Unequal Load–Source Impedances 28
3.9 Impedance Matching and Frequency Compensation of Semiconductor Devices (RF and Microwave FETs) 29
3.10 Real Frequency Impedance Matching Using Hilbert Transforms and No Optimization 30
3.11 The Ideal Transformer as the Ideal Impedance Matching Device Deficiencies of the Conventional Transformer 32
3.12 Transmission Line Impedance Transformers–Guanella and Ruthroff 34
References 35

4 Automated Impedance Matching Network Design Process and Design Examples with SPICE Performance Evaluation 37
4.1 Introduction 37
4.2 "L" Lumped Element Impedance Transformer – Automated Design and SPICE Performance Evaluation 37
4.3 Single Transmission Line Segment Impedance Transformer – Automated Design and SPICE Performance Evaluation 43
4.4 Single/Double Closed-/Open-Circuit Transmission Line Stub Impedance Transformer – Automated Design and SPICE Performance Analysis 47
4.5 "Convert" Complex Load to Resistance – Parallel/Series Capacitor or Inductor Load Reactance – Automated Design and SPICE Performance Analysis 50
4.6 Active Impedance Matching with Bi-junction Transistors in Feedback Loop and Shorted Transmission Line Segments to Neutralize Reactances 60
4.7 Broad/Wide Band Impedance Matching with Unequal Source and Load Resistance Norton Single Element Series/Shunt Transforms – Automated Design and SPICE Performance Analysis 66

4.8 Broad/Wide Band Impedance Matching with Unequal Source and Load Resistance General Norton to Element PI, T Transforms – Automated Design and SPICE Performance Analysis 71
4.8.1 Key Observation 74
4.9 Broad/Wide Band Low-Pass Matching Network – Equal Source–Load Resistance – Parallel Load Capacitor – One-End Reflections – Automated Design and SPICE Performance Analysis 76
4.10 Low-Pass Configuration Matching Network Design with Bode–Fano Constraints – Automated Design and Evaluation of Continued Fraction Form of Driving Impedance Function 81
4.11 Non-optimized Real Frequency Method – Automated Design and Numerical Integration Evaluation of Hilbert Transform 85
4.12 Guanella 1:*n* Ideal and Lossy Transmission Line Transformer Impedance Matching – Real Load Automated Design and SPICE Performance Analysis 87
4.13 Guanella 1:*n* Ideal and Lossy Transmission Line Transformer Impedance Matching Complex Series Load – Automated Design and SPICE Performance Analysis 92
4.14 Guanella 1:*n* Ideal and Lossy Transmission Line Transformer Impedance Matching – Complex Parallel RL Load – Automated Design and SPICE Performance Analysis 96
4.15 Simple Broad/Wide Band Impedance Matching Scheme 100

Appendix A: Supplied C Computer Language Executables for the Linux and Windows Operating Systems 103

Appendix B: How to Install MingW (GCC Compiler Suite) on Computers Running Windows Operating System 105

Index 107

Chapter 1
Introduction and Problem Statement

1.1 Introduction

Impedance matching is a technique to guarantee that maximum/optimum signal power is transferred from the signal source to the receiving device to ensure minimum signal power reflection back to the source. No electronic signal processing circuit (especially those operating at 100 s of MHz or 10s of GHz—e.g., telecommunication/wireless communication equipment, consumer electronic devices) can operate without impedance matching between its subcircuits. There are two types of impedance matching—broad/wide and narrow bands. Of these two, broad/wide band impedance matching is more challenging to implement because maximum/optimum signal power must be transferred from the source to the load over a predefined band of frequencies. Although both broad and narrow band impedance matching are based on the *conjugate matching* criterion, broad band impedance matching has additional restrictions on bandwidth and gain, imposed by the Bode–Fano inequalities, which arise because impedance matching circuits include reactive components (capacitors/inductors), whose reactances are frequency dependent. Therefore, although traditional broadband impedance matching subcircuit design calculations start out with the Bode–Fano inequalities, for real-world design cases, very clever techniques have been formulated to neutralize reactive components, thereby forcing a constant gain over the operating frequency range. Each of these very interesting topics will be examined in the subsequent chapters, followed by a chapter dedicated to design examples using C computer language executable modules that implement the complicated design calculations for both broad and narrow band impedance matching schemes.

A. Banerjee, *Automated Broad and Narrow Band Impedance Matching for RF and Microwave Circuits*, https://doi.org/10.1007/978-3-319-99001-9_1

1.2 Why Impedance Matching Is Essential

Impedance matching is essential because signal reflection occurs due to impedance mismatch in the signal path. Consider two ideal transmission line segments with characteristic impedances Z_1 and Z_2 driving a load Z_L (Fig. 1.1a). The wave equations for current and voltage at the interface of the two transmission line segments are first derived and then related to the impedance. Using Kirchoff's current and voltage laws on the "loop" formed by the two conductors in Fig. 1.1b gives:

$$V(x) - V(x+dx) - Ldx\left(\frac{\partial I}{\partial t}\right) = 0 \text{ or } \frac{-\partial V}{\partial x} = L\left(\frac{\partial I}{\partial t}\right) \tag{1.1}$$

The charge dQ accumulating in length dx in time dt is $dQ = CdxdV$ in terms of the change of voltage dV between the two conductors and the capacitance per unit length of the transmission line. That is,

$$Q = CdVdx = (I(x) - I(x+dx))dt \text{ or } \frac{-\partial I}{\partial x} = C\left(\frac{\partial V}{\partial t}\right) \tag{1.2}$$

Combining (1.1) and (1.2), the wave equations for current and voltage are:

$$\frac{\partial^2 I}{\partial x^2} = LC\frac{\partial^2 I}{\partial t^2} \text{ and } \frac{\partial^2 V}{\partial x^2} = LC\frac{\partial^2 V}{\partial t^2} \tag{1.3}$$

The solutions to (1.3) are plane waves:

$$I_{\text{plus}}\mathrm{e}^{j(kx-wt)}, \quad I_{\text{minus}}\mathrm{e}^{-j(kx+wt)} \text{ and } V_{\text{plus}}\mathrm{e}^{j(kx-wt)}, \quad V_{\text{minus}}\mathrm{e}^{-j(kx+wt)} \tag{1.4}$$

where plus/minus indicate wave propagation in the forward/reverse directions. The wave velocity is:

$$v = \frac{w}{k} = \frac{1}{\sqrt{LC}}$$

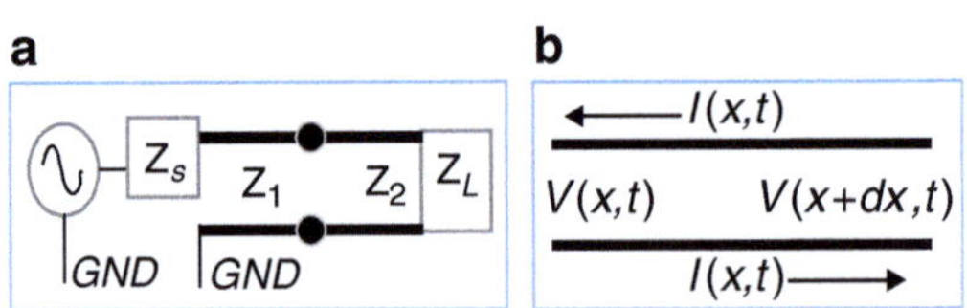

Fig. 1.1 (**a**) Signal source, two transmission (**b**) Unit length of transmission line segment and source and load impedances with currents and voltages indicated

where C and L are the capacitance and inductance per unit length, respectively. Then combining (1.4) with (1.2)–(1.3):

$$V_{\text{plus}} = \sqrt{\frac{L}{C}} I_{\text{plus}} = Z I_{\text{plus}}, V_{\text{minus}} = -\sqrt{\frac{L}{C}} I_{\text{minus}} = -Z I_{\text{minus}} \quad (1.5)$$

Therefore, the average power carried by the forward going or reverse going waves is:

$$P_{\text{plus}} = \frac{V_{\text{plus}}^2}{2Z} \text{ and } P_{\text{minus}} = \frac{-V_{\text{minus}}^2}{2Z} \quad (1.6)$$

Now consider the interface between the two transmission line segments with different characteristic impedances Z_1 and Z_2. The incident current/voltage wave is split at the interface into two—transmitted and reflected waves. Therefore, the current and voltage waves propagating along the second transmission line segment are:

$$I_{2,\text{plus}} = I_{1,\text{plus}} + I_{1,\text{minus}} \text{ and } V_{2,\text{plus}} = V_{1,\text{plus}} + V_{1,\text{minus}} \quad (1.7)$$

where the subscripts minus and plus denote the reflected and transmitted waves, respectively. Then $I_{2,\text{plus}}\left(\frac{Z_2}{Z_1}\right) = I_{1,\text{plus}} - I_{1,\text{minus}}$, and after combining and solving these expressions, $I_{1,\text{minus}} = \left(\frac{Z_1 - Z_2}{Z_1 + Z_2}\right) I_{1,\text{plus}}$ and $I_{2,\text{plus}} = \left(\frac{2Z_1}{Z_1 + Z_2}\right) I_{1,\text{plus}}$. So whenever

$$Z_1 \neq Z_2 \; I_{1,\text{minus}} \neq 0 \quad (1.8)$$

The expression (1.8) *embodies the key reason why impedance matching has to be implemented—or else there will be reflected signals.*

The discussion so far has not considered the load impedance, which in most cases is complex (parallel/series RC/RL). As both a capacitor and an inductor are reactive, their reactance (effectively resistance) varies with frequency. In case Z_1 and Z_2 are complex, the reflected and transmitted signal powers are, respectively:

$$P_{1,\text{minus}} = \left|\frac{Z_1 - Z_2}{Z_1 + Z_2}\right|^2 . P_{1,\text{plus}} \text{ and } P_{2,\text{plus}} = \frac{4 \cdot Z_1 \cdot Z_2}{|Z_1 + Z_2|^2} . P_{1,\text{plus}} \quad (1.9)$$

The first one of the pair of eqs. (1.8) can be rewritten as $I_{\text{reflected}} = I_{1,\text{minus}} = \frac{1 - \frac{Z_2}{Z_1}}{1 + \frac{Z_2}{Z_1}} I_{1,\text{plus}} = -\text{gamma} I_{\text{incident}}$, where the reflection coefficient is

$$\text{gamma} = \frac{\frac{Z_2}{Z_1} - 1}{\frac{Z_2}{Z_1} + 1} = \frac{z - 1}{z + 1} \quad (1.10)$$

and the complex position-dependent impedance is $Z(x) = Z_1\left(\frac{1+\text{gammafunc}\,\mathrm{e}^{-2.j.\text{beta}.x}}{1-\text{gammafunc}\,\mathrm{e}^{-2.j.\text{beta}.x}}\right)$, where beta is the wave number defined as $\frac{6.28}{\text{wavelength}}$. The position-dependent impedance will be real if e^{-2jkx} is real and then the corresponding length of the transmission line segment is

$$x = \frac{(\text{wavelength})}{12.56}\arctan\left(\left(\frac{2Z_1X_2}{Z_1^2 - R_2^2 - X_2^2}\right)\right) \text{ and, } Z_2 = R_2 + jX_2 \tag{1.11a}$$

As x must be less than 0,

$$x = \frac{(\text{wavelength})}{12.56}\arctan\left(\left(\frac{2Z_1X_2}{-Z_1^2 + R_2^2 + X_2^2}\right)\right) \tag{1.11b}$$

If load reactance $X_2 = 0$, $x = 0$ is the position closest to the load at which $Z(x)$ is real.

1.3 How to Implement Impedance Matching: Conjugate Matching Condition

The most general case of a signal source with an impedance Z_S driving a load impedance Z_L through a transmission line segment of characteristic impedance Z_0 is shown in Fig. 1.2. Both Z_S and Z_L are complex, i.e., $Z_S = R_S + jX_S$ and $Z_L = R_L + jX_L$. The input impedance looking onto the loaded transmission line from the signal source end is:

$$Z_{\text{input}} = \frac{Z_L + j.Z_0.\tan(\text{beta}.l)}{Z_0 + j.Z_L.\tan(\text{beta}.l)},$$

where beta is the signal wave number and l is the physical length of the transmission line segment. The reflection coefficient of the load is $\text{gamma}_L = \frac{Z_L - Z_0}{Z_L + Z_0}$. The voltage wave traveling down the line is

$$V(x) = V_S^{\text{plus}}\left(\mathrm{e}^{-j.\text{beta}.z} + \text{gamma}_{L}.\mathrm{e}^{j.\text{beta}.x}\right) \tag{1.12}$$

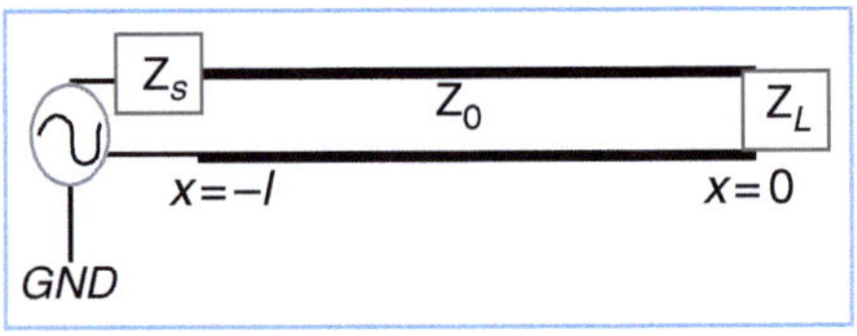

Fig. 1.2 Signal source with complex impedance driving complex load

Using

$$V_S^{plus} = \left(\frac{V_S Z_0}{Z_S + Z_0}\right)\left(\frac{e^{-j.beta.l}}{1 - gamma_L.gamma_G.e^{-2j.beta.l}}\right), \quad gamma_G = \frac{Z_G - Z_0}{Z_G + Z_0}, \text{ and}$$

$$SWR = \frac{1 - |gamma_L|}{1 + |gamma_L|} \tag{1.13}$$

where SWR is the standing wave ratio, the signal power delivered to the load is:

$$P = \frac{\Re\left(V_{input}.I_{input}^2\right)}{2} = \frac{V_{input}^2}{2}\cdot\Re\left[\frac{1}{Z_{input}}\right] = \frac{V_G^2}{2}\cdot\left|\frac{Z_{input}}{Z_{input}+Z_S}\right|^2\cdot\Re\left[\frac{1}{Z_{input}}\right] \tag{1.14a}$$

Using $Z_S = R_S + jX_S$ and $Z_L = R_L + jX_L$, (1.14a) is reduced to:

$$P = \frac{R_{input}|V_S|^2}{(R_{input}+R_S)^2 + (X_{input}+X_S)^2} \tag{1.14b}$$

To maximize the power delivered to the load, the partial derivatives of (1.14b) with respect to R_{input} and X_{input} are computed and set to 0:

$$\frac{\partial P}{\partial R_{input}} = 0 = \frac{1}{(R_{input}+R_S)^2 + (X_{input}+X_S)^2} + \frac{-2R_{input}(R_{input}+R_S)}{\left((R_{input}+R_S)^2 + (X_{input}+X_S)^2\right)^2} \tag{1.15a}$$

$$\text{or } R_S^2 - R_{input}^2 + (X_{input}+X_S)^2 = 0 \tag{1.15b}$$

Similarly, from the other partial derivative,

$$\frac{\partial P}{\partial X_{input}} = 0 = \frac{-2R_{input}(X_{input}+X_S)}{\left((R_{input}+R_S)^2 + (X_{input}+X_S)^2\right)^2} \text{ or } X_{input}(X_{input}+X_S)0 \tag{1.16}$$

The eqs. (1.15b) and (1.16) are satisfied if and only if.

$$R_S = R_L \text{ and } X_S = -X_L \tag{1.17}$$

This is the *conjugate matching* condition and forms the basis for broad and narrow band impedance matching. For broad band impedance matching, additional Bode–Fano inequalities impose conditions on gain and bandwidth. These will be examined in later chapters.

Exercises

- Derive each of the three equations in (1.13).

Chapter 2
Narrow Band Impedance Matching

2.1 Introduction

This chapter examines in detail narrow band impedance matching, i.e., how to maximize signal power delivery from a signal source to a load, given that both the source and load impedances are complex, at a specified frequency with very narrow and tight upper and lower side bands. Narrow band impedance matching is based entirely on conjugate matching, examined in the previous chapter.

2.2 The Common *L* Impedance Transformer with Complex Load and Source Impedances

The popular *L* impedance matching circuit or impedance transformer is analyzed in detail, and key design equations are derived. Both source and load impedances are complex, i.e., $Z_S = R_S + jX_S$, $Z_L = R_L + jX_L$, $X_S \neq 0$, and $X_L \neq 0$. The derivation of the design equations are based on the conjugate matching conditions. With reference to Fig. 2.1a, the admittance is:

$$\frac{1}{Z_S} = jB + \frac{1}{R_L + j(X + X_L)} \tag{2.1a}$$

which after some manipulation yields:

$$\begin{aligned} R_L + j(X + X_L) = {} & [R_S(1 - B(X + X_L)) - BR_L X_S] \\ & + j[BR_S R_L + X_S - BX_S(X + X_L)] \end{aligned} \tag{2.1b}$$

Now equating the real and imaginary parts of (2.1b), one obtains:

A. Banerjee, *Automated Broad and Narrow Band Impedance Matching for RF and Microwave Circuits*, https://doi.org/10.1007/978-3-319-99001-9_2

Fig. 2.1 (**a**, **b**) Two common *L* impedance transformers with complex source–load impedances

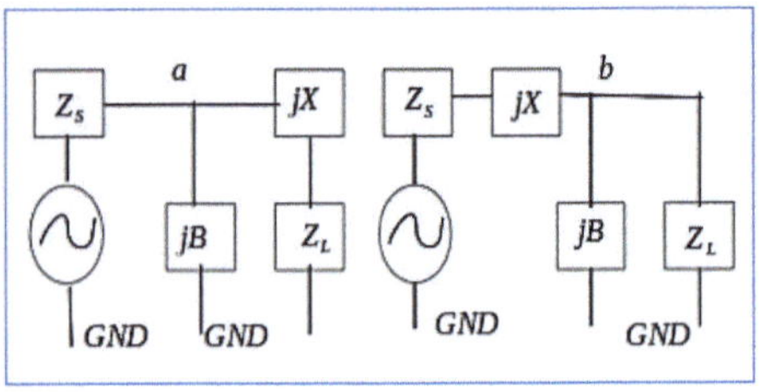

$$R_L = R_S - B[R_S(X + X_L) - R_L X_S] \tag{2.1c}$$

and

$$(X + X_L) = BR_S R_L + X_S(1 - B(X + X_L)) \tag{2.1d}$$

After some manipulation of both Eqs. (2.1c) and (2.1d), the expression for X is:

$$X = \left(\frac{1}{R_S}\right)\left[\frac{R_S - R_L}{B} - (R_S X_L + R_L X_S)\right] \tag{2.1e}$$

Substituting (2.1e) in (2.1d) and using the intermediate expressions $a = R_S - R_L$ and $b = R_S X_L + R_L X_S$ generate, after some manipulation, the following quadratic equation for B:

$$B^2(X_L X_S - R_L R_S - bX_S) + B\left(\frac{aX_S}{R_S} - b + X_L - X_S\right) + \frac{a}{R_S} = 0 \tag{2.1f}$$

This quadratic equation can have real roots if and only if

$$\left(\frac{aX_S}{R_S} - b + X_L - X_S\right)^2 \geq 4(X_L X_S - R_L R_S - bX_s)\left(\frac{a}{R_S}\right) \tag{2.1g}$$

If the inequality (2.1f) is satisfied, the values of B are calculated by solving Eq. (2.1f). These values of B are substituted into Eq. (2.1e) to obtain the values of X. This is left as an exercise for the reader.

Now consider the impedance matching circuit of Fig. 2.1b. The impedance equation is:

$$Z_S = jX + \frac{1}{jB + \frac{1}{R_L + jX_L}} \tag{2.1h}$$

This expression, after expansion and rearrangement, yields:

$$Z_S[(1 - BX_L) + jBR_L] = -XBR_L + R_L + jX(1 - BX_L) \tag{2.1i}$$

Equating the real and imaginary parts of (2.1i) gives:

$$X(1 - BX_L) = BR_LZ_S - X_L \text{ and } B(XR_L - X_LZ_S) = R_L - Z_S \tag{2.1j}$$

Now substituting $Z_S = R_S + jX_S$ in (2.1i) and then simplifying, rearranging and equating the real and imaginary parts of both sides,:

$$R_S(1 - BX_L) + BR_LX_S = R_L(1 - BX) \quad \text{and} \quad B(X_SX_L + R_SR_L) - X_S = X_L + X - BXX_L \tag{2.1k}$$

The first expression in (2.1k), $R_S(1 - BX_L) + BR_LX_S = R_L(1 - BX)$, after some manipulation, gives

$$X = \frac{R_S - R_L}{BR_L} - \left(\frac{R_L}{R_S}\right)X_S + \left(\frac{R_S}{R_L}\right)X_L \tag{2.1l}$$

After (2.1l) is substituted in the second expression in (2.1k), some manipulation/rearrangement yields the following quadratic equation for B:

$$B^2\left[(X_LX_S + R_LR_S) + \left(\frac{R_S}{R_L}\right)X_L^2 - \left(\frac{R_L}{R_S}\right)X_LX_S\right] + B\left[\frac{R_LX_S}{R_S} - \frac{R_SX_L}{R_L} - (X_L + X_S)\right] - \left(\frac{R_L - R_S}{R_L}\right) = 0.$$

Real values of B can be obtained from this quadratic equation if its determinant is ≤ 0. These real values of B can be back substituted to provide the corresponding values of X.

2.3 Single Transmission Line Segment to Match Complex Load–Source Impedances

A signal source with impedance $Z_S = R_S + jX_S$ drives a load impedance $Z_L = R_L + jX_L$ through transmission line segment with characteristic impedance Z_0 and electrical length theta = beta.$l = \frac{6.281}{\text{lambda}}$, where lambda is the signal wavelength and l the physical length of the transmission line segment. The input impedance of the series combination of the transmission line segment and the load is:

$$Z_{\text{INP}} = \frac{Z_0[R_L + j(X_L) + Z_0 \tan(\text{beta}.l)]}{Z_0 - X_L \tan(\text{beta}.l) + jR_L \tan(\text{beta}.l)]} \tag{2.2a}$$

From the conjugate matching condition, the complex conjugate of the source impedance equals the combined input impedance of the series transmission line segment and the load impedance. So from (2.2a),

$$R_S - jX_S = \frac{Z_0[R_L + j(X_L) + Z_0 \tan(\text{beta}.l)]}{Z_0 - X_L \tan(\text{beta}.l) + jR_L \tan(\text{beta}.l)} \tag{2.2b}$$

After expanding, rearranging and equating real and imaginary parts from both sides, one gets:

$$Z_0(R_S - R_L) + X_L \tan(\text{beta}.l)(R_L - R_S) = 0 \tag{2.2c}$$

and

$$R_L R_S \tan(\text{beta}.l) - 2X_L Z_0 + \left(X_L^2 - Z_0^2\right) \tan(\text{beta}.l) = 0 \tag{2.2d}$$

These are two equations in two unknowns, Z_0 and theta. Solving them gives:

$$Z_0 = \sqrt{\left(\frac{R_S\left(R_L^2 + X_L^2\right) - R_L\left(R_S^2 + X_S^2\right)}{R_L - R_S}\right)}$$
$$\text{theta} = \arctan\left(\frac{Z_0(R_S - R_L)}{R_S X_L - R_L X_S}\right) \tag{2.2e}$$

So a transmission line segment with characteristic impedance and electrical length, as specified in (2.2e), will match the two impedances $Z_S = R_S + jX_S$ and $Z_L = R_L + jX_L$.

2.4 The Concept of Reflection Coefficients

Consider a simple setup consisting of a signal source with a source impedance $Z_S = R_S + jX_S$ driving a load $Z_L = R_L + jX_L$. The amplitude of the signal is $2V_{\text{INP}}$. Using the voltage divider formula, $V_{\text{OUT}} = 2V_{\text{INP}}\left(\frac{Z_L}{Z_L + Z_S}\right)$. As $Z_S \neq Z_L$, the reflected voltage is $V_{\text{REF}} = V_{\text{OUT}} - V_{\text{INP}} = V_{\text{INP}}\left[\frac{Z_L - Z_S}{Z_L + Z_S}\right]$. Then the reflection coefficient is defined as

$$\text{Reflection Coefficient} = \text{Gamma} = \frac{V_{\text{REF}}}{V_{\text{INP}}} = \frac{Z_{\text{L}} - Z_{\text{S}}}{Z_{\text{L}} + Z_{\text{S}}} \tag{2.2f}$$

2.5 The Lambda/12 Transmission Line Impedance Transformer with Complex Load–Source Impedances

Two consecutively connected transmission line segments with characteristic impedance Z_2, Z_i and physical lengths l_2,l_1 connect to a load $Z_{\text{L}} = R_{\text{L}} + jX_{\text{L}}$, i.e., the load is at $x = l_1 + l_2$. The goal is to eliminate reflected waves in the transmission line segment Z_1, l_1. So $V_{1,\text{ plus}} = I_{1,\text{ plus}}Z_1$ $V_{2,\text{ plus}} = I_{2,\text{ plus}}Z_2$, $V'_{1,\text{plus}} = I'_{1,\text{plus}}Z_1$, and $V'_{1,\text{minus}} = I'_{1,\text{minus}}Z_1$, where the primed values denote current and voltage in the transmission line segment Z_1, l_1.

From continuity considerations, $I_{1,\text{ plus}} = I_{2,\text{ plus}} + I_{2,\text{ minus}}$ and $V_{1,\text{ plus}} = V_{2,\text{ plus}} + V_{2,\text{ minus}}$. Now substituting voltages with currents,

$$\begin{aligned} I_{1,\text{plus}}\left(\frac{Z_1}{Z_2}\right) &= I_{2,\text{plus}} = I_{2,\text{minus}} \text{ or } I_{2,\text{plus}} = \frac{Z_2 + Z_1}{2Z_2} I_{1,\text{plus}}, I_{2,\text{minus}} \\ &= \frac{Z_2 + Z_1}{2Z_2} I_{1,\text{minus}} I_{2,plus} = \frac{Z_2 + Z_1}{2Z_2} I_{1,plus} \end{aligned} \tag{2.3a}$$

Using identical continuity arguments at the interface $x = l_2$ gives:

$$I_{2,\text{plus}}\text{e}^{j\text{beta}l_2} + I_{2,\text{minus}}\text{e}^{-j\text{beta}l_2} = I'_{1,\text{plus}}\text{e}^{j\text{beta}l_2} + I'_{1,\text{minus}}\text{e}^{-j\text{beta}l_2} \tag{2.3b}$$

and

$$v_{2,\text{plus}}\text{e}^{j\text{beta}l_2} + v_{2,\text{minus}}\text{e}^{-j\text{beta}l_2} = v'_{1,\text{plus}}\text{e}^{j\text{beta}l_2} + v'_{1,\text{minus}}\text{e}^{-j\text{beta}l_2} \tag{2.3c}$$

From these two Eqs. (2.3a) and (2.3b), voltages are eliminated, giving:

$$\left(I_{2,\text{plus}}\text{e}^{j\text{beta}l_2} - I_{2,\text{minus}}\text{e}^{-j\text{beta}l_2}\right)\left(\frac{Z_2}{Z_1}\right) = I'_{1,\text{plus}}\text{e}^{j\text{beta}l_2} + I'_{1,\text{minus}}\text{e}^{-j\text{beta}l_2} \tag{2.3d}$$

Then solving (2.3b) and (2.3d) gives: for primed forward and reflected currents:

$$I'_{1,\text{plus}} = \left[\frac{(Z_1 + Z_2)^2 = (Z_1 - Z_2)^2\text{e}^{-2j\text{beta}l_2}}{4Z_1Z_2}\right] I,\text{plus} \tag{2.3e}$$

and

$$I'_{1,\text{minus}} = \left[\frac{(Z_1^2 - Z_2^2)}{4Z_1Z_2}\right]\left(e^{-2j\text{netal}_2} - 1\right) I_1, \text{plus} \tag{2.3f}$$

Identical current and voltage continuity conditions are applied to the junction $x = l_1 + l_2$, and this gives:

$$\begin{aligned} I &= I'_{1,\text{plus}} e^{j\text{beta}l_2(l_1+l_2)} + I'_{1,\text{minus}} e^{-j\text{beta}l(l_1+l_2)} \text{ and} \\ V &= V'_{1,\text{plus}} e^{j\text{beta}l(l_1+l_2)} + V'_{1,\text{minus}} e^{-j\text{beta}(l_1+l_2)} \end{aligned} \tag{2.3g}$$

Then eliminating voltages with currents (substituting (2.3e) and (2.3f) in (2.3g)) gives

$$I = \frac{\left((Z_1 + Z_2)^2 A - (Z_1 - Z_2)^2 B + \left(Z_1^2 - Z_2^2\right)(C - D)\right) I_{1,\text{plus}}}{4Z_1Z_2} \tag{2.3h}$$

where $A = e^{j\ \text{beta}(l_1+l_2)}$, $B = e^{j\ \text{beta}(l_1-l_2)}$, $C = e^{-j\ \text{beta}(l_1-l_2)}$ and $D = e^{-j\ \text{beta}(l_1+l_2)}$

Also,

$$I = \frac{\left((Z_1 + Z_2)^2 A - (Z_1 - Z_2)^2 B + \left(Z_1^2 - Z_2^2\right)(C - D)\right) I_{1,\text{plus}}}{4ZZ_2} \tag{2.3i}$$

Now equating (2.3h) to (2.3i) gives, after some manipulation and rearrangement:

$$\begin{aligned} (Z_1 + Z_2)^2 \cos(\text{beta}(l_1 + l_2)) = {} & \left(Z_1^2 + Z_2^2\right)\cos(\text{beta}(l_1 - l_2)) \\ & + jZ_1Z_2 \sin(\text{beta}(l_1 - l_2)) \end{aligned}$$

As the left hand side of the above expression is real, it must be the case that $l_1 = l_2$, $(Z = Z_2)$. Then

$$\cos(2.\text{beta}.1) = \frac{Z_1^2 + Z_2^2}{(Z_1 + Z_2)^2}, Z = Z_2 \tag{2.3j}$$

From (2.3j), when $Z_1 \approx Z_2, l = \frac{\text{lamda}}{12}$, where lambda is the signal wavelength. Extending this analysis a little further, it can be shown that the quarter wave transmission line impedance transformer, used to match real loads, follows from the above analysis. This is left as an exercise for the reader.

2.6 Single Series or Shunt Closed/Open Transmission Line Impedance Transformers

A transmission line stub of length l is placed at a distance d from a load $Z_L = R_L + jX_L$. The characteristic impedance of each of the transmission line feeding the signal, the transmission line segment of length l, and the stub is Z_0. The transmission line stub may be closed or open circuited and can be in parallel to or in series with the load. First, the parallel/shunt stub case is examined. The impedance of the line of length *dm* seen from the load is [1, 2]:

$$Z = Z_0\left[\frac{(R_L + jX_L) + jZ_0 \tan(\text{beta}.d)}{Z_0 + j(R_L + jX_L)\tan(\text{beta}.d)}\right] \tag{2.4a}$$

Using $t =$ tan (beta.d) and admittance $Y = \frac{1}{Z}$ (2.4a) becomes:

$$Y = \frac{Z_0 + j(R_L + jX_L)t}{Z_0R_L + jZ_0(X_L + Z_0t)} = G + jB \tag{2.4b}$$

Simplifying (2.4b), the expressions for the conductance and susceptance are:

$$G = \frac{R_L(1+t^2)}{R_L^2 + (X_L + Z_0t)^2} \text{ and } B = \frac{R_L^2t + (Z_0 - R_Lt)(X_L + Z_0t)}{Z_0\left[R_L^2 + (X_L + Z_0t)^2\right]} \tag{2.4c}$$

Now, d is selected such that $G = Y_0 = \frac{1}{Z_0}$, and then the first expression in (2.4c) becomes a quadratic equation in t, whose two roots are:

$$t = \frac{X_L + \sqrt{\frac{R_L}{Z_0}\left[(Z_0 - R_L)^2 + X_L^2\right]}}{(R_L - Z_0)}, R_L > Z_0 \text{ and}$$
$$t = \frac{X_L - \sqrt{\frac{R_L}{Z_0}\left[(Z_0 - R_L)^2 + X_L^2\right]}}{(R_L - Z_0)}, R_L > Z_0 \tag{2.4d}$$

When $R_L \approx Z_0\, t = \frac{-X_L}{2Z_0}$ and then $\frac{d}{\text{lamda}} = \frac{\arctan(t)}{6.28}, t \geqslant 0$ and

$$\frac{d}{\text{lambda}} = \frac{3.14 + \arctan(t)}{6.28},\ t \geqslant 0 \tag{2.4e}$$

These calculated values of t are used to compute the values of susceptance B and then the values of the open- and short-circuited stub lengths are:

$$\frac{l_{\text{open}}}{\text{lambda}} = \left(\frac{-1}{6.28}\right) \arctan\left(\frac{B}{Y_0}\right) \text{ and } \frac{l_{\text{closed}}}{\text{lambda}} = \left(\frac{1}{6.28}\right) \arctan\left(\frac{Y_0}{B}\right) \quad (2.4f)$$

For single series closed- or open-circuited stubs, the analysis proceeds as above. The admittance of the subcircuit looking from the load into the transmission line segment of length d and the series (closed/open) stub is:

$$Y = Y_0\left[\frac{(G_L + jB_L) + jtY_0}{Y_0 + jt(G_L + jB_L)}\right], \text{ where } t = \tan(\text{beta}.d), Y_0 = \frac{1}{Z_0} \quad (2.4g)$$

Using $Z = R + jX = \frac{1}{Y}$ $R = \frac{G_L(1+t^2)}{G_L^2 + (B_L + Y_0 t)^2}$ and

$X = \frac{G_L^2 t - (Y_0 - B_L t)(B_L + Y_0 t)}{Y_0\left[G_L^2 + (B_L + Y_0 t)^2\right]}$.

Then using identical reasoning as before, $\frac{d}{\text{lambda}} = \frac{\arctan(t)}{6.28}, t \geqslant 0$ and $\frac{d}{\text{lambda}} = \frac{3.14 + \arctan(t)}{6.28}, t \geqslant 0$ and

$$\frac{l_{\text{closed}}}{\text{lambda}} = \left(\frac{-1}{6.28}\right) \arctan\left(\frac{X}{Z_0}\right) \text{ and } \frac{l_{\text{open}}}{\text{lambda}} = \left(\frac{1}{6.28}\right) \arctan\left(\frac{Z_0}{X}\right) \quad (2.4h)$$

The intermediate derivation steps are left as an exercise for the reader.

2.7 Double Series or Shunt Closed/Open Transmission Line Impedance Transformer

Two closed- or open-circuited (or open/closed circuit combination) transmission line stubs of physical lengths l_1, l_2 are connected to the two ends of a transmission line segment of physical length d with a load $Z_L = R_L + jX_L$ as in Fig. 2.2. The characteristic impedances of each of the feed transmission line and the transmission line segments and stub is Z_0. All analysis starts from the load end. The admittance to the left of the first stub is $Y_1 = G_L + j(B_L + B_1)$, where B_L and B_1 are respectively the conductance and susceptance of the load and first stub respectively. Similarly, the admittance to the right of the second stub is [2]:

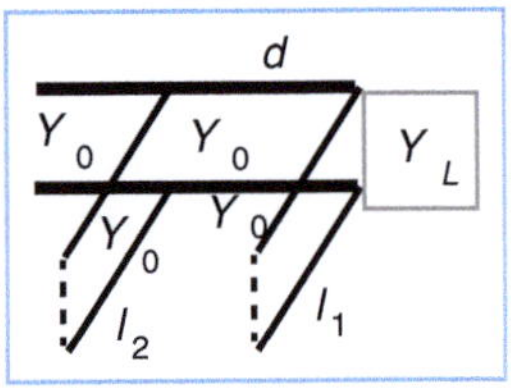

Fig. 2.2 Double closed/ open stub transmission line impedance transformer. Dotted lines indicate that stubs may be closed or open circuited

$$Y_2 = Y_0\left[\frac{G_L + j(B_L + B_1 + Y_0 \tan(\text{beta}.d))}{Y_0 + j\tan(\text{beta}.d)(G_L + j(B_L + B_1))}\right] \tag{2.5a}$$

Applying the substitutions $t = \tan(\text{beta}.d)$ and $Y_0 = \frac{1}{Z_0}$ and recognizing that the real part of (2.5a) must equal Y_0 generate a quadratic equation for G_L, whose roots are

$$G_L = Y_0\left(\frac{1+t^2}{2t^2}\right)\left[1 + \sqrt{1 - \frac{4t^2(Y_0 - B_L t - B_1 t)^2}{Y_0^2(1+t^2)^2}}\right] \text{ and}$$
$$G_L = Y_0\left(\frac{1+t^2}{2t^2}\right)\left[1 - \sqrt{1 - \left(\frac{4t^2(Y_0 - B_L t - B_1 t)^2}{Y_0^2(1+t^2)^2}\right)}\right] \tag{2.5b}$$

As G_L is always real, it must be that

$$0 \le \frac{4t^2(Y_0 - t(B_1 + B_L))^2}{Y_2^0(1+t^2)^2} < 1 \tag{2.5c}$$

The inequality (2.5c) implies that

$$0 \le G_L \le \frac{Y_0}{(\sin(\text{beta}.d))^2} \tag{2.5d}$$

The inequality (2.5d) specifies the range of possible values of G_L for a predefined value of d. Then the first stub susceptance can be determined from the quadratic equation for G_L. These are:

$$B_1 = -B_L + \frac{Y_0 + \sqrt{(1+t^2)G_L Y_0 - (G_L t)^2}}{t} \text{ and}$$
$$B_1 = -B_L + \frac{Y_0 - \sqrt{(1+t^2)G_L Y_0 - (G_L t)^2}}{t} \tag{2.5e}$$

Similarly, from (2.5a), the values of the susceptance B_2 are:

$$B_2 = \frac{Y_0\sqrt{\left(Y_0 G_L\left(1 + t^2 - (G_L t)^2\right)\right)} + G_L Y_0}{G_L t} \text{ and}$$
$$B_2 = \frac{-Y_0\sqrt{\left(Y_0 - G_L\left(1 + t^2 - (G_L t)^2\right)\right)} + G_L Y_0}{G_L t} \tag{2.5f}$$

Now given that the two stubs can be open or closed circuited, their lengths are

$$\frac{l_{\text{open}}}{\text{lambda}} = \left(\frac{1}{6.28}\right)\arctan\left(\frac{B}{Y_0}\right), B = B_1, B_2 \text{ and } \frac{l_{\text{closed}}}{\text{lambda}} = -\left(\frac{1}{6.28}\right)\arctan\left(\frac{Y_0}{B}\right), B = B_1, B_2 \tag{2.5g}$$

Exercises

- Derive Eq. (2.1e) and (2.1f) starting with Eqs. (2.1c) and (2.1d).
- Derive Eqs. (2.2c) and (2.2d) starting with Eq. (2.3b).
- Derive Eqs. (2.3b) and (2.3c) and use them to obtain Eqs. (2.3d), (2.3e), and (2.3f).
- Analyze and explain how the quarter wave impedance transformer, used to match real sources and loads, can be derived from the lambda by 12 transmission line transformer.

References

1. Grebennikov A (2004) RF and microwave power amplifier design. The McGraw Hill Company ISBN-13: 978-0071828628; ISBN-10 0071828621
2. Pozar DM (2011) Microwave engineering, 4th edn. John Wiley and Sons, Inc. ISBN-13 978-0-470-63155-3; ISBN-10 0470631554

Chapter 3
Broadband Impedance Matching

3.1 Introduction

Broadband impedance matching addresses the very important problem of how to maximize signal power delivery from a complex source to a complex load over a large frequency range while maintaining a constant gain (within acceptable tolerance limits). A complex load has a nonzero reactance, so the impedance varies as a function of the signal frequency and therefore the gain. So along with the conjugate matching principle, broadband impedance matching is constrained by the Bode–Fano inequalities, which impose tight theoretical bounds on the gain and bandwidth of a broadband impedance matching network. But in practice, it is very difficult to design a broadband impedance matching network that strictly adheres to the Bode–Fano constraints. A viable and widely used alternative is to cancel out the effects of the reactive elements, making the load and/or source appear as a pure resistance—the Zobel network and active impedance matching. These very interesting topics, as well as related (and very important) topics, such as the Norton impedance transformers, are examined in detail in this chapter.

3.2 Bode–Fano Conditions on Gain Bandwidth of Broadband Impedance Matching Networks

Although broadband impedance matching exploits the same conjugate matching condition as that of the narrow band impedance matching, there is another key criterion that must be satisfied. This arises from two contradictory requirements that the designer has to reconcile:

- Reflection coefficient must be minimized over the maximum possible bandwidth

A. Banerjee, *Automated Broad and Narrow Band Impedance Matching for RF and Microwave Circuits*, https://doi.org/10.1007/978-3-319-99001-9_3

Fig. 3.1 Canonical impedances used in Bode–Fano inequalities

This translates to how to minimize the number of matching network stages to satisfy a given matching bandwidth. These essential requirements are estimated from the Bode–Fano [1, 2] criterion that specifies the minimum reflection coefficient that can be achieved with predefined canonical load impedances (parallel and/or series resistor-capacitor or resistor-inductor) (Fig. 3.1). This theoretical limit on the reflection coefficient is:

$$\int \ln\left(\left[\frac{1}{|\text{ gamma}(w)\ |}\right]\right) dw \leq \frac{3.14}{\text{tau}},\ \ 0 \leq w \leq \infty \tag{3.2a}$$

where tau is the time constant of the complex load, gamma the reflection coefficient, and w the angular frequency in radians/second. For a series RC load, the Bode–Fano integral is:

$$\int \left(\frac{1}{w^2}\right) \ln\left(\left[\frac{1}{|\text{ gamma}(w)\ |}\right]\right) dw \leq 3.14.\text{tau},\ \ 0 \leq w \leq \infty \tag{3.2b}$$

The mathematical relationship expressed by (3.2b) is shown in Fig. 3.1, which shows the ideal reflection coefficient over the specified bandwidth. Clearly, a larger bandwidth corresponds to higher value of reflection coefficient. Therefore, if the absolute value of the reflection coefficient is set to a specific value within a given bandwidth, and equals 1 outside this frequency band, then from (3.2a):

$$\int \ln\left(\left[\frac{1}{|\text{ gamma}(w)\ |}\right]\right) dw = (w_2 - w_1) \ln\left(\frac{1}{|\text{ gamma }|}\right) \leq \frac{3.14}{\text{tau}}\quad w_1 \leq w \leq \infty \tag{3.2c}$$

which implies that

$$|\text{gamma}_{\text{min}}| = \exp\left(\frac{-3.14}{(w_2 - w_1)\text{tau}}\right) \tag{3.2d}$$

Then from (3.2c) and the series RC and parallel RL load impedances:

$$|\text{gamma}|_{\text{min}} = \exp\left(\frac{-3.14\text{w}_0^2 \cdot \text{tau}}{w_2 - w_1}\right) \tag{3.2e}$$

where $w_0 = \sqrt{w_2 w_1}$ is the band center frequency. ***It is extremely important to note that the Bode–Fano theoretical limits on gain and bandwidth can be achieved only and only if there are an infinite number of sections in the matching network.*** In addition, the frequency bandwidth with minimum reflection coefficient magnitude is achieved with a loaded quality factor $Q_L = w_0 \cdot \text{tau}$ for the series RL and parallel RC load impedances and $Q_L = \frac{1}{w_0 \cdot \text{tau}}$ for the series RC and parallel RL load impedances. As these conditions are difficult to realize in practice, best approximations are used—a Chebyschev matching network with finite number of sections, whose pass-band ripple is equal to the minimum reflection coefficient. In such cases, the minimum reflection coefficient is $|\text{gamma}|_{\min} = \exp\left(\frac{-3.14 Q_0}{Q_L}\right)$, where $Q_0 = \frac{w_0}{w_2 - w_1}$.

3.3 Converting Complex Series RC or RL or Parallel RC or RL Load into Pure Resistance: The Zobel Network and Parallel to Equivalent Series Impedance Transformation

Though it is essential to take into account gain and bandwidth in the design and implementation of broadband impedance matching networks for complex loads, using the Bode–Fano inequalities as the starting point in the design calculations is not fruitful as the calculated Bode–Fano limits apply only to a matching network with an infinite number of sections. Most importantly, the last several design calculations are very complicated, involving an evaluation of a continued fraction expansion of ratios of high-order polynomials. Though specialized techniques as the Brune's method exist, they are mostly manual and very difficult to automate, thereby highly error prone and time consuming.

The constant resistance network (also called 1 pole diplexer or impedance flattening network or officially the Zobel [3, 4] network) adroitly overcomes this problem. *This is achieved by "converting" a complex series RC or RL impedance into a pure resistor,* i.e., *effectively canceling out the reactance. A parallel RC or RL impedance can be easily transformed into an equivalent series RC or RL impedance, followed by constructing the Zobel network for the transformed equivalent series RC or RL impedance.* As the parallel to series transformation calculation requires a frequency value, the band center frequency is the ideal candidate, and most importantly, as the Zobel network neutralizes all reactances, frequencies do not matter in the final design. *So the Zobel network is truly a broadband impedance matching device.*

The Zobel network's usefulness and versatility is because its input impedance is fixed at design time, completely independent of the transfer function. The design equations, derived from balanced bridge conditions, are very simple. With reference to Fig. 3.2a, note that impedance Z_B ***does not exist*** *in the real world. Zobel network*

Fig. 3.2 (**a–c**) Zobel network as a balanced bridge, series RL load, and equivalent Zobel network

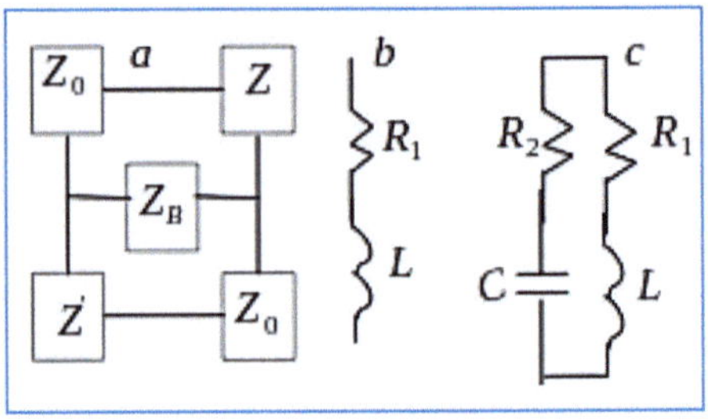

Z_B *is only for ease of deriving the equations.* So from the balanced condition, $\frac{Z}{Z_0} = \frac{Z_0}{Z'}$, and the input impedance is:

$$\frac{1}{Z_{INPUT}} = \frac{1}{Z + Z_0} + \frac{1}{Z' + Z_0} \tag{3.3a}$$

So from the balanced condition,

$$Z_{INPUT} = Z_0 \tag{3.3b}$$

For a series RC or RL impedance, $Z_0 = R$, ***which implies that the input impedance is independent of the input signal frequency***. In addition, Z' and Z are duals of each other, i.e., if Z is a capacitor, Z' is an inductor and vice versa.

The expressions derived above can also be obtained from Kirchoff's current law (KCL). Consider the simple series RL load (Fig. 3.2b) connected to an ideal V Volt step-function voltage source. If the load was purely reactive, the current through the inductor would rise exponentially to a final value. On the other hand, if the load were purely resistive, then the current would also rise in a step function to a final steady value. With the series RL circuit, the time constant is $\text{tau}_1 = \frac{L}{R_1}$. In order that the voltage source supply a step current, a parallel branch is needed that compensates for the slow rising load current I_L. This is provided by the series R_2C connected in parallel to the series R_1L. The current through the R_2C leg I_C is instantaneously equal to $\frac{V}{R_2}$ and decreases exponentially to zero with a time constant $\text{tau}_2 = R_2C$. However, the time it takes to charge the capacitor (current drops exponentially to zero) must equal the time it takes for the inductor to get energized, i.e., $\text{tau}_1 = \text{tau}_2$. This is possible if and only if

$$R_1 = R_2 \quad \text{and} \quad R_2C = \frac{L}{R_1} \tag{3.3c}$$

Although the above analysis is based on series RC, RL impedances, a parallel RC or RL pair can be easily transformed into an equivalent series RC or RL pair, and then the Zobel network can be constructed for this transformed series RC or RL. The reader is encouraged to examine the Zobel network based design examples in Chap. 4.

The explicit formulas for converting parallel RL impedance to equivalent series RL and vice versa are:

$$L_{\mathrm{SER}} = \frac{L_{\mathrm{PAR}} \cdot Q^2}{1+Q^2} \text{ and } R_{\mathrm{SER}} = \frac{R_{\mathrm{PAR}}}{1+Q^2}$$

where $Q = \frac{R_{\mathrm{PAR}}}{w_0 L_{\mathrm{PAR}}} = \frac{w_0 L_{\mathrm{SER}}}{R_{\mathrm{SER}}}$ and w_0 is the band center frequency. When converting from series to equivalent parallel RL impedance,

$$L_{\mathrm{PAR}} = \frac{L_{\mathrm{SER}}\left(1+Q^2\right)}{Q^2} \text{ and } R_{\mathrm{PAR}} = R_{\mathrm{SER}}\left(1+Q^2\right) \tag{3.3d}$$

Identical formulas hold for parallel RC to equivalent series RC and vice versa. The concept of frequency is immaterial for a Zobel network as the effects of all reactive circuit elements are canceled out.

3.4 Active Impedance Matching and Neutralizing Reactances or Susceptances of Complex Impedances or Admittances

The Bode–Fano inequalities arise from the frequency dependency of the reactances of both the capacitor and the inductor and impose strict thresholds on the bandwidth and gain. The technique to *neutralize* the frequency dependency of a complex load is very simple and straightforward, although the implementation is complicated.

Let a complex load have the admittance $Y = G + j \cdot B$, where G is the conductance and B the susceptance. ***Theoretically, to neutralize B, a susceptance of opposite polarity, i.e, –j·B, must be attached across this admittance as admittances add in parallel.*** As negative susceptances or reactances do not exist in the real world, a negative susceptance or impedance is a meta device that has to be constructed out of existing capacitors, inductors, transmission line segments, and sometimes semiconductor devices. So negative admittances/impedances that work by using properties of semiconductor devices are called *active* negative capacitors or inductors.

To understand how a transmission line segment can be used to neutralize a reactance, consider the impedance of a transmission line of length L, with a load $Z_L = R_L + jX_L$ attached at the end,

$$Z_{\mathrm{INP}}(-L) = Z_0 \left[\frac{Z_L + jZ_0 \tan\,(\mathrm{beta}.L)}{Z_0 + jZ_L \tan\,(\mathrm{beta}.L)}\right] \tag{3.4a}$$

Now, instead of the load attached to one end, if the end is short circuited, the corresponding admittance and impedances are respectively

$$Y_{\mathrm{INP}} = \frac{-j\cot(\mathrm{beta}.L)}{Z_0}, Z_{\mathrm{INP}} = jZ_0 \tan(\mathrm{beta}.L) \tag{3.4b}$$

To nullify a reactance, a shorted transmission line segment of characteristic impedance Z_0 is connected in parallel to the complex load impedance. Rewriting the load impedance in terms of the admittance, $Y_L = G_L + jB_L$, and then to neutralize the load susceptance, $\frac{-\cot(\mathrm{beta}.L)}{Z_0} = B_L$, or equivalently,

$$\frac{-\cot(\mathrm{beta}.L)}{Z_0} = \frac{-X_L}{R_L^2 + X_L^2} \tag{3.4c}$$

Now, using the band center frequency and the known values of Z_0, R_L, X_L, the value of the inductance, the physical length L of the short-circuited transmission line segment is easily computed. The impedance $Z_L = R_L + jX_L$ now appears as a pure resistor to the signal source.

Exploiting similar concepts, Linvill [18] and Perry [19] proposed *using subcircuits with active (semiconductor) devices (high-frequency bi-junction transistors or field-effect transistors (FETs)) that convert a reactive element into its negative counterpart.*

The idea of negative reactance is novel. Consider a capacitor with a negative capacitance value, controlled by the equation $V(t) = \left(\frac{-1}{C}\right)\int I(t)\,dt$, where *the negative sign represents an intrinsic property that causes the current flowing through the capacitor to flow in the reverse direction (compared to current flowing through a nonnegative capacitor) or, equivalently, the voltage-time gradient to be negative*. Such a negative capacitance when attached in series to a positive capacitor causes the combination to act as a short circuit, and when connected in parallel, the combination acts as an open circuit. Both of these results can be easily derived from expressions of parallel and series combinations of capacitors and are left as an exercise for the reader.

When placed in series with a resistor and a positive capacitor, the combination acts as a pure resistor $Z = R + \frac{j}{wC} - \frac{j}{wC}$ provided that the positive and negative capacitors are equal in magnitude. When connected in parallel to a series resistor positive capacitor pair, the combination has an admittance given by $Y = \frac{Rw^2C_\mathrm{p}^2}{1 + (RwC_\mathrm{p})^2} + \frac{jwC_\mathrm{p}}{1 + (RwC_\mathrm{p})^2} - jwC_\mathrm{n}$, where C_p and C_n are the positive and negative capacitances, respectively. Given that at frequency range of 100 s of MHz to 10s of GHz, $(RwC_\mathrm{p})^2 < 1$, the expression for the admittance becomes $Y \approx \left((wC_\mathrm{p})^2\left(R + j\left(\frac{1}{wC_\mathrm{p}} - \frac{1}{wC_\mathrm{n}}\right)\right)\right)$, and in the case $C_\mathrm{p} \approx C_\mathrm{n}\, Y \approx R(wC_\mathrm{p})^2$, Y is a real valued frequency-dependent number. Similar arguments may be applied to the analysis of a negative inductor, and this is left as an exercise for the reader. Armed with these basic concepts, the design and implementation of a negative impedance converter is examined in detail.

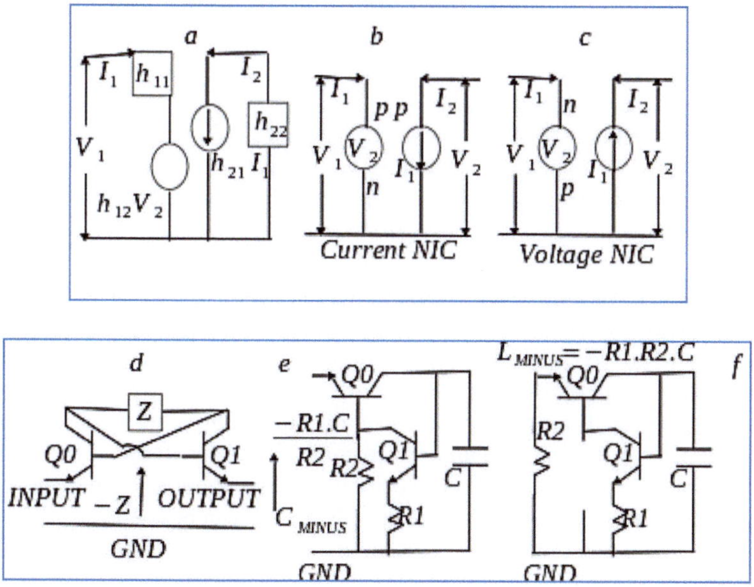

Fig. 3.3 (**a**) represents the h parameter model of a negative converter. (**b** and **c**) represent, respectively, the current and voltage NICs. (**d**, **e**, **f**) Floating negative impedance converter and grounded negative capacitor and inductor, respectively

A negative impedance converter is a two-port device having the novel property that an immittance shunted across one of terminals will appear as the negative valued immittance at the other pair of terminals. An analogy is an ideal transformer with a transformation ratio of –K:1. With reference to Fig. 3.3a, the following two equations that relate the current and voltage at the two ports, in terms of h parameters, are:

$$V_1 = h_{11}I_1 + h_{12}V_2, I_2 = h_{21}I_1 + h_{22}V_2 \tag{3.4d}$$

So if a load impedance Z_L is placed across the terminals of the NIC, the input impedance may be written as (using 3.4a)

$$Z_{\text{INP}} = \frac{V_1}{I_1} = h_{11} = \frac{h_{12}h_{21}Z_L}{1 + h_{22}Z_L} \tag{3.4e}$$

The product $h_{12}h_{21}$ is called the *devices transmission gain* or *devices conversion ratio* and is denoted as K. Now, for $Z_{\text{INP}} = -Z_L$ to hold, $h_{11} = h_{22} = 0, h_{12} = h_{21} = 1$. From these equations, two types of negative impedance converters, depending on whether current or voltage inversion occurs, can be implemented. A NIC that inverts current *does not* change the input/output voltage polarity, and the complimentary conditions hold for the voltage-inverting NIC. These two types of negative impedance converters are given in Fig. 3.3b, c, respectively.

Practical negative impedance converters are augmented versions of the original designs of Linvill and occur in two varieties: floating and grounded (Fig. 3.3d–f). These are used in active RC filters, devices that enhance the Q values of passive resonators and, most importantly, *broadband impedance matching of electrically small antennas.* As each of these subcircuits uses either RF or microwave bi-junction transistors or FETs, impedance matching with negative impedance converters is called *active impedance matching. A major drawback of this scheme is that the designer must ensure that the selected transistor's operating frequency range matches the frequency range over which the impedance matching is to be achieved.* Of the two negative impedance converter types (floating, grounded), the floating type is more common as a series element in impedance matching circuits, where it neutralizes the positive reactance. Negative impedance converters are open-circuit stable at one port and short-circuit stable at the other. This means that with a passive impedance connected across port 2, the network formed by open circuiting port 1 is stable. Similarly, with a passive impedance connected at port 1, the network formed by short circuiting port 2 is stable. The conditional stability of a negative impedance converter restricts the magnitude of the impedance that can be connected to both the short- and open-circuit stable ports. That is, should Z_{L1}, Z_{L2} be, respectively, the load impedances connected to ports 1, 2 of a negative impedance converter, then the following inequalities must be satisfied:

$|Z_{LI}| \geqslant |Z_{\mathrm{INP},\,1}|$ (open-circuit stable port) and $|Z_{L2}| \leqslant |Z_{\mathrm{INP},\,2}|$ (short-circuit stable port), where $Z_{\mathrm{INP},\,1}$, $Z_{\mathrm{INP},\,2}$ are, respectively, the impedances looking into ports 1, 2.

When a negative impedance converter is part of an impedance matching network, it must be properly terminated, and the natural frequencies of the network must lie in the left half of the complex plane. Most importantly, the feedback loop must provide both the gain and phase margin for stability.

As mentioned earlier, a negative impedance converter makes a positive reactance appear as negative reactance to the remainder of the network it is connected to. So a capacitor is modeled as a capacitor (negative) in parallel with a conductance G (positive or negative). Similarly, an inductor is a series combination of a negative inductor and resistor (positive or negative).

3.5 Low-Pass Configuration Matching Network Design with Bode–Fano Constraints and Table Lookup

Ladder configuration low-pass (shunt capacitor–series inductor) networks can be used to impedance match loads as in Fig. 3.3a, starting with the Bode–Fano inequalities [5], which give:

$$\int \ln\left(\frac{1}{|\,\mathrm{rho}_1\,|}\right) dw = \frac{3.14}{L_{\mathrm{LOAD}}} - 3.14 \sum P_{rl}\; 0 \leq w \leq \infty \tag{3.5a}$$

$$\int w^2 \ln\left(\frac{1}{|\,\mathrm{rho}_1\,|}\right) dw = \frac{3.14}{L_{\mathrm{LOAD}}^3}\left(\frac{L_{\mathrm{LOAD}}}{C_{\mathrm{LOAD}}} - 0.33\right) + 1.04\sum P_{rl}^2\ 0 \le w \le \infty \tag{3.5b}$$

where rho_1 is the reflection coefficient function to the matching network P_{rl}, its zeros in the right half of the complex plane, and w the angular frequency. From (3.5a), the area under the curve $\ln\left(\frac{1}{|\mathrm{rho}_1|}\right)$ is a constant, with some additive constants (zeros of the reflection function in the right half plane). Same conclusions may be drawn from (3.4b), and both equations represent gain-bandwidth type of relationships.

When the matching is constructed/synthesized as a Chebyschev-type low-pass ladder network, the transfer function is of the form:

$$\frac{P_{\mathrm{available}}}{P_{\mathrm{out}}} = 1 + \mathrm{epsilon}\left(\cosh\left(n' \operatorname{arccosh}(w')\right)\right)^2\ 1 \le w' \le \infty \tag{3.5c}$$

$$\frac{P_{\mathrm{available}}}{P_{\mathrm{out}}} = 1 + \mathrm{epsilon}\left(\cos\left(n' \arccos(w')\right)\right)^2\ 0 \le w' \le 1 \tag{3.5d}$$

where epsilon is the Chebyschev pass-band ripple factor, w' the angular frequency normalized to cut off frequency of $w_0 = 1$, and n' the number of reactive elements in the network under analysis. The expression w' is defined as $w' = \frac{2\left(w^2 - w_0^2\right)}{\left(w_{\mathrm{U}}^2 - w_{\mathrm{L}}^2\right)}$, $w_0 = \sqrt{\frac{w_{\mathrm{U}}^2 + w_{\mathrm{L}}^2}{2}}$, and w_{U} and w_{L} are, respectively, the upper and lower limits of the operating frequency band. The mean operating frequency w_{m} is the arithmetic mean of w_{U} and w_{L} scaled so that $w_{\mathrm{m}} = 1$. The relative bandwidth is $w = w_{\mathrm{U}} - w_{\mathrm{L}}$, where $w_{\mathrm{L}} = 1 - \frac{w}{2}$ and $w_{\mathrm{U}} = 1 + \frac{w}{2}$.

These simple frequency normalizations are used to simplify (3.5a), (3.5b), introducing a new parameter delta that is typically obtained from table lookup. The next key step is to evaluate the poles and zeros of the reflection coefficient function rho_1. First, the poles and zeros of the conventional Chebyschev low-pass transfer function $\left|\mathrm{rho}_1'\right|^2$ are evaluated. The poles are:

$$\begin{gathered} P_{\mathrm{p}}' = \sinh\left(a + \frac{3.14 \cdot j}{n'}\left(m + \frac{1}{2}\right)\right) n' \text{even and } P_{\mathrm{p}}' = \sinh\left(-a - \frac{3.14 \cdot j}{n'}\left(m + \frac{1}{2}\right)\right) n' \text{even} \\ P_{\mathrm{p}}' = \sinh\left(a + \frac{3.14 \cdot j \cdot m}{n'}\right) n' \text{odd and } P_p' = \sinh\left(-a - \frac{3.14 \cdot j \cdot m}{n'}\right) n' \text{odd} \end{gathered} \tag{3.5e}$$

where the two parameters a, b are defined as:

$$(\sin \mathrm{h}(n'b))^2 = \frac{\text{delta}}{\text{epsilon}} \text{ and } (\sin \mathrm{h}(n'a))^2 = \frac{1 + \text{delta}}{\text{epsilon}} \tag{3.5f}$$

The expressions for the zeros are evaluated with b replacing a in the above equations.

The poles and zeros of the mapped reflection coefficient function are evaluated from the above expressions for poles and zeros as

$$p' = \frac{-2j(p^2 + w_0^2)}{w_U^2 - w_L^2} \tag{3.5g}$$

obtained by making the substitutions $w' = -jp'$ and $w = -jp$ in $w' = \frac{2(w^2 - w_0^2)}{(w_U^2 - w_L^2)}$.

The gain-bandwidth integrals can be evaluated and then plotted for various values of n to obtain a nomogram used to look up values of delta. Once the values of the poles and zeros are calculated using (3.5g), the reflection coefficient function $\text{rho}_1(p)$ is easily evaluated. The matching network input impedance is $Z(p) = \frac{1 - \text{rho}_1}{1 + \text{rho}_1}$. To synthesize the matching network, the impedance expression is expanded as a continued function expansion. The resulting component values are all normalized and need to be both frequency and impedance scaled.

This scheme is not very amenable to automation because nomogram or table lookup requires judicious choice of tolerances as both the values being searched for and the table values are all floating point values, ruling out exact matches. Evaluation of the reflection coefficient function, followed by the matching network impedance function, from the computed values of poles and zeros can only be partially automated. Most importantly, automated evaluation of continued fraction expansion is difficult and is best done with specialized mathematical software packages such as IMSL (International Mathematical Standard Library). Another scheme [6] for designing and synthesizing impedance matching networks for complex loads starts with the Bode–Fano inequalities and then relies on a two-stage table lookup to complete the design calculations. This makes the scheme inappropriate for automation as the accuracy and reliability of the extracted answers depend entirely on how judiciously search tolerances are chosen.

3.6 Impedance Matching of Equal Real Source and Load Resistors with Load Having Nonzero Reactance Low-Pass and Band-Pass Matching Network

Sometimes source and load impedances have identical real part (source and load resistances are the same), but the reactances might have different values and types. Specifically, the source is real and the load is complex, with values of the source and load resistances being identical. The intermediate impedance matching network can be low pass or band pass. If the load impedance is parallel RC, then a low-pass configuration and shunt capacitor at the output of value C or a little higher would cancel out the load capacitor. The shunt capacitor in the low-pass matching network is reduced by the load capacitor. From the Bode–Fano inequalities, the cut-off frequency of the low-pass matching network is reduced to $f_{\max} = \frac{g}{6.28.\text{bandwidth}.R.C_{\text{shunt}}}$, where g is the low-pass prototype coefficient or g value of the matching network element adjacent to the load capacitor. The bandwidth is taken to be 1. Similarly, for a series resistor-inductor load, the maximum frequency is $f_{\max} = \frac{gR}{6.28.\text{bandwidth}.L_{\text{series}}}$.

A band-pass-type impedance matching network can tackle series RC loads, as well as parallel RL, in addition to the shunt RC and series RL. For the shunt RL load, the maximum frequency is $f_{\max} = \frac{\text{bandwidth}.R}{6.28.g.L_{\text{shunt}}}$, and the corresponding maximum frequency for the series RC load is $f_{\max} = \frac{\text{bandwidth}}{6.28.g.R.C_{\text{series}}}$.

A far more accurate, effective and powerful solution is the Zobel network. The Zobel network is a natural fit for the series RC or RL load impedance, and for the shunt RC or RL load impedance, the equivalent series RC or RL load can be constructed, followed by the Zobel network for the transformed RC or RL load impedance. With the Zobel network, all frequency-related issues theoretically vanish.

3.7 Norton Series and Shunt Transforms for Matching Unequal Load Source Impedances

The Norton subnetworks or transforms (series, shunt, and generalized) [7] are a set of extremely simple yet versatile techniques to tackle a variety of real-world design issues, e.g., matching of unequal source–load impedances, replacing too large or too small-valued ladder configuration filter components (capacitors, inductors), with physically realizable values and so on. The simplest two of these transforms are the series and shunt transforms, when a reactive element is replaced by an equivalent PI or T subnetwork and an ideal transformer. An ideal transformer is a perfect impedance transformer, to be examined in detail later. In this case, the turn ratio of the ideal transformer can be a positive fraction. The series and shunt Norton transforms are shown in Fig. 3.4. The governing equations for the series and shunt transforms are respectively:

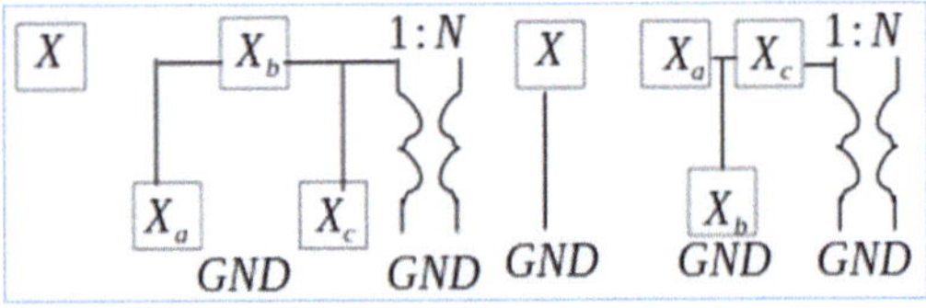

Fig. 3.4 Series (PI) and shunt (T) Norton transforms or subnetworks

$$X_a = \frac{X}{1-N} X_b = \frac{X}{N} X_c = \frac{X}{N(N-1)}, \text{ and } X_a = \frac{(N-1)X}{X} X_b = \frac{X}{N} X_c = \frac{(1-N)X}{N} \tag{3.7a}$$

Noteworthy is that some of the PI or T component values might be negative, but they can be tackled easily. To understand the power and utility of these two simple transforms, the reader must examine the relevant design examples in Chap. 4. Both the series and shunt Norton transforms may be applied as many times as the designer so wishes at any location inside the ladder network to be transformed.

3.8 Generalized Norton Transforms for Matching Unequal Load–Source Impedances

The four canonical generalized Norton transform subcircuits are shown in Fig. 3.5a. As ideal transformers cannot be realized in practice, especially at frequencies in the range of 100 s of MHz to 10 s of GHz, each combination of two reactive elements and ideal transformer is replaced by either a PI or T network, consisting entirely of reactive elements, as shown in Fig. 3.5b. Contrary to claims, the Norton transform is not a truly broadband solution as the final subcircuit consists entirely of reactive elements, and unlike the Zobel subnetwork, there is no mechanism to cancel out the frequency dependency of the reactive elements. Thus, the responsibility of tackling the frequency-dependent effects of capacitors and inductors rests with the designer. The prime candidate to achieve this goal is the Zobel subnetwork, which is an ideal device to cancel out the frequency-dependent effects of complex source or load impedances, allowing the Norton transformer to match unequal source–load resistances.

As ideal transformers cannot be fabricated in practice, each of the four canonical Norton subnetworks in Fig. 3.5a has equivalent PI and T subcircuit forms, consisting entirely of reactive elements. The equivalent PI and T equivalent subnetworks, when the canonical Norton subnetwork consists of two capacitors and the ideal transformer, is shown in Fig. 3.5b. The equivalent PI and T subnetworks, when the canonical Norton subnetwork consists of two inductors and the ideal transformer, are very similar to those shown in Fig. 3.5b.

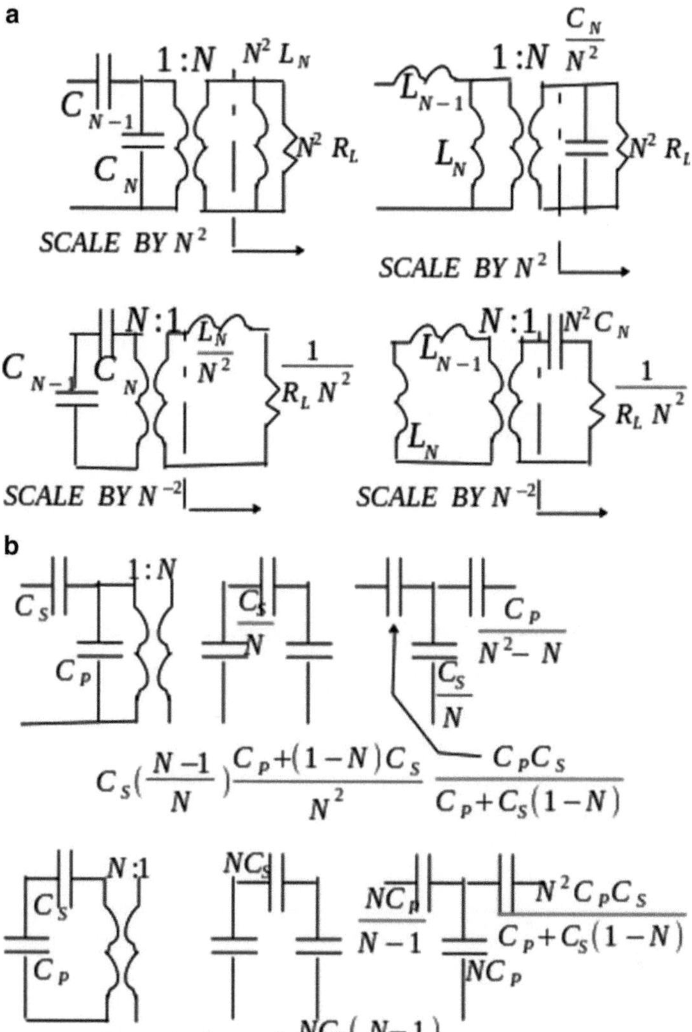

Fig. 3.5 (**a**) Four canonical Norton transform subcircuits. (**b**) Equivalent PI and T subnetworks when the canonical Norton subnetwork consists of two capacitors and an ideal transformer

3.9 Impedance Matching and Frequency Compensation of Semiconductor Devices (RF and Microwave FETs)

A simple lumped element model of an RF and/or microwave field-effect transistor (FET) is shown in Fig. 3.6. The gate impedance is series RC, and the output is a parallel RC. To guarantee that an amplifier using RF/microwave FET maintains uniform gain over a wide operating frequency range, both the capacitors for the input

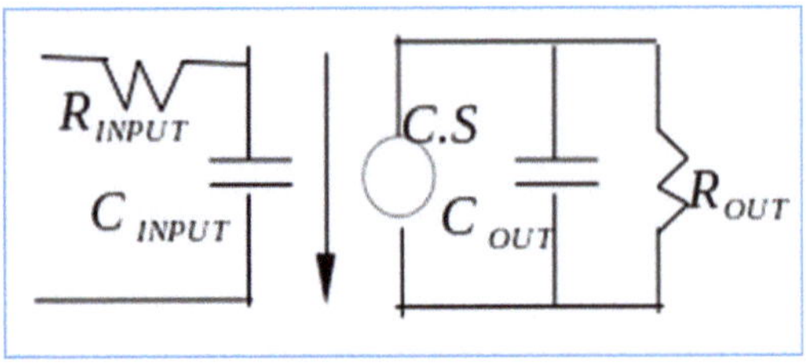

Fig. 3.6 Simple RF/microwave FET model. C.S is the current source, and the arrow indicates the direction of current flow

and output should be canceled out, i.e., both the input and output impedances must be purely resistive. The input (gate) impedance can be converted into a resistor by constructing the corresponding Zobel subnetwork for it. The output parallel RC can be easily converted into the equivalent series RC impedance, and then the Zobel subnetwork for that transformed RC impedance can be constructed. With both input and output impedances "converted" into pure resistances, the gain is uniform over the entire operating frequency range.

3.10 Real Frequency Impedance Matching Using Hilbert Transforms and No Optimization

Real frequency technique [9–15], in its original form, relied on a set of optimizations on large data sets from impedance match to complex impedances. The drawback to this approach is that most optimization schemes do not converge if the matrices on which they are operating are ill-conditioned. Therefore, a simplified real frequency method has been formulated that addresses this problem. However, it does involve calculation of Hilbert transform of large data sets as an intermediate step, as well as the *conjugate gradient* method as one of the last calculation steps. The simplified scheme is examined in detail now. Consider a signal source with a source impedance driving a load through the matching network (Fig. 3.7a). To synthesize the matching network, the impedance function of the matching network must be determined. *The key input data set is the set of real and imaginary parts of the load impedance, measured over the operating frequency range.*

The transducer power gain (TPG) of the circuit in Fig. 3.7 is, using $Z_2 = R_2 + jX_2$ and $Z_L = R_L + jX_L$:

$$\mathrm{TPG} = \frac{4R_2R_L}{(R_2 + R_L)^2 + (X_2 + X_L)^2} \tag{3.10a}$$

Assuming that the matching network consists of a pure resistive part and a pure reactive part, the matching network impedance can be rewritten as:

Fig. 3.7 Template network for nonoptimized real frequency scheme analysis

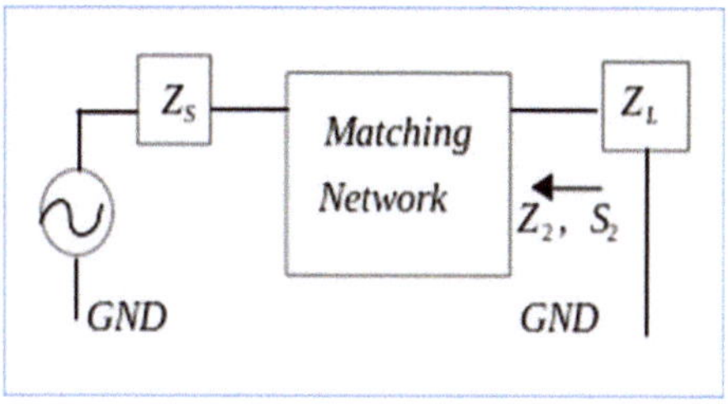

$$Z_2 = Z_{\text{MINIRESISTANCE}} + Z_{\text{FOSTER}} = R_{\text{MINIRESISTANCE}} + j(X_{\text{MINIRESISTANCE}} + X_{\text{FOSTER}}) \tag{3.10b}$$

Then using the substitutions $X_2 = X_{\text{MINIRESISTANCE}} + X_{\text{FOSTER}}$ $R_2 = R_{\text{MINIRESISTANCE}}$, (3.10a) becomes

$$\text{TPG} = \frac{4R_{\text{MINIRESISTANCE}}R_L}{(R_{\text{MINIRESISTANCE}} + R_L)^2 + (X_{\text{MINIRESISTANCE}} + X_{\text{FOSTER}} + X_L)^2} \tag{3.10c}$$

Then from the conjugate matching condition:

$$R_2 = R_{\text{MINIRESISTANCE}} = R_L \; X_2 = X_{\text{MINIRESISTANCE}} + X_{\text{FOSTER}} = -X_L \tag{3.10d}$$

Now, another simplifying assumption is made, i.e., only the imaginary part is matched as per the conjugate matching condition so that

$$\text{TPG} = \frac{4R_{\text{MINIRESISTANCE}}R_L}{(R_{\text{MINIRESISTANCE}} + R_L)^2} \tag{3.10e}$$

By simplifying (3.10e), the expression for the ratio of $R_{\text{MINIRESISTANCE}}$ to R_2, alpha is defined as:

$$\text{alpha} = \frac{R_{\text{MINIRESISTANCE}}}{R_{\text{L}}} = \frac{R_2}{R_{\text{L}}} = \frac{2 - \text{TPG} \pm \sqrt{1 - \text{TPG}}}{\text{TPG}} \tag{3.10f}$$

Then $R_2 = R_{\text{MINIRESISTANCE}} = \text{alpha} \cdot R_{\text{L}}$, and once R_2 is calculated, then $X_{\text{MINIRESISTANCE}}$ is computed as

$$X_{\text{MINIRESISTANCE}} = \text{HilbertTransform}(R_{\text{MINIRESISTANCE}}) \tag{3.10g}$$

At this stage, a suitable TPG function has to be selected, and the most common one is the Chebyschev as $\text{TPG} = \frac{1}{1 + (\text{epsilon}.T_n(w))^3}$, where epsilon is the pass-band ripple factor and $T_n(w)$ is the nth order Chebyschev polynomial. The value of the Chebyschev polynomial fluctuates in about a steady value. This steady-state value is called TPG_{flat} and has been evaluated analytically for some canonical loads such as the series RC, for which

$$TPG_{flat} = 1 - e^{\frac{-6.28}{RCw_c}}, \quad (3.10h)$$

where w_c is the cut-off frequency

Then the pass-band ripple may be expressed as $\text{epsilon}^2 = \frac{1 - TPG_{flat}}{TPG_{flat}}$. Combining all this information, the nonoptimized real frequency method is select or calculate TPG_{flat}. This is most often supplied by the designer.

Using the value for TPG_{flat}, compute epsilon:

- Select the order of the Chebyschev polynomial, and then using the above computed value of epsilon, compute TPG values.
- Use the values for TPG to compute the values for alpha.
- Using the designer-supplied values for the real and imaginary parts of the load impedance over the operating frequency range, compute the real part of the minimum reactive matching network.
- Compute the Hilbert transform of the real part of the minimum reactive matching network (previous step), and then compute the imaginary part of the same network.
- Compute the Foster network data, and then use the conjugate gradient technique to model the minimum reactive network and the Foster network.
- Use the results of the modeling task to synthesize the matching network and compute component values.

Clearly, the real frequency technique is very compute intensive and requires significant effort to extract meaningful results.

3.11 The Ideal Transformer as the Ideal Impedance Matching Device Deficiencies of the Conventional Transformer

Let L_1, L_2 and M be the coil inductances and mutual inductance of the two coils 1, 2 of an ideal transformer. Then the magnetic flux linked with a single turn of coil 1 is $\text{phi}_0 = \int B_0 dA = L_0 I_1$, where B_0 is the flux density due to one turn, L_0 is the self-inductance of one turn, and I_1 is the current through coil 1. If coil 1 has N_1 turns, then the total magnetic flux due to coil 1 is $N_1 B_0$ and the total flux linked by coil 1 is $\text{phi}_1 = N_1^2 L_0 I_1 = L_1 I_1$. Identical arguments hold for coil 2. As this is an ideal transformer, all magnetic flux generated by coil 1 is linked to coil 2 and vice versa: $\text{phi}_{12} = N_1 N_2 L_0 I_1 = MI_1$ and $\text{phi}_{21} = N_1 N_2 L_0 I_2 = MI_2$. So $L_1 = N_1^2 L_0$, $L_2 = N_2^2 L_0$, and $M = N_1 N_2 L_0$ or $M^2 = L_1 L_2$.

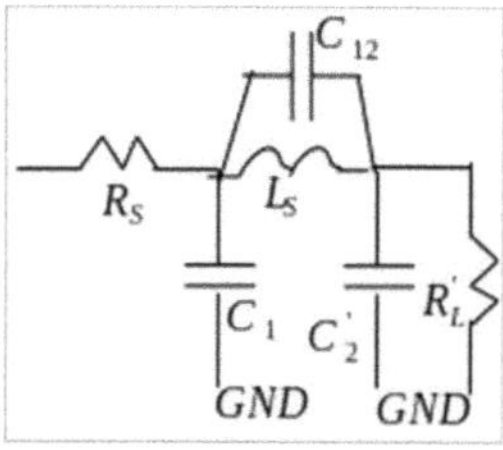

Fig. 3.8 High-frequency model of a conventional transformer

As this is an ideal transformer, energy conservation in the two coils indicate $I_1 V_1 = I_2 V_2$ and $N_1 V_2 = N_2 V_1$ so that if the source and load impedances are $Z_1 Z_2$, respectively, $I_1 = \frac{V_1}{Z_1}$ and $I_2 = \frac{V_2}{Z_2}$. So from the energy balance equation, $\frac{V_1^2}{Z_1} = \frac{V_2^2}{Z_2}$ or $\frac{V_1^2}{V_2^2} = \frac{Z_1}{Z_2}$. Also, $\frac{V_1^2}{V_2^2} = \frac{N_1^2}{N_2^2}$ so that

$$\frac{N_1}{N_2} = \sqrt{\frac{Z_1}{Z_2}} \tag{3.11}$$

As $N_1 N_2$ are real, $Z_1 Z_2$ must be real too. This does not mean that the ideal transformer cannot be used with complex loads. The complex load can be "converted" into a pure resistor, as elaborated on in the previous sections.

This ideal transformer may be approximated closely by a carefully constructed ferrite core transformer. Conventional soft iron sheet core transformers fail at high frequencies, the reasons being (Fig. 3.8):

- Inductor reactances are directly proportional to frequency so that as frequency increases, the current through the primary and secondary decreases, in turn decreasing the flux linked from the primary to the secondary.
- Increased coil resistances cause increased ohmic loss.
- Both leakage inductances of the primary and leakage inductances of the secondary with respect to the primary increase.
- Both parasitic capacitance of the primary and parasitic capacitance of the secondary with reference to the primary increase with reference to the operating frequency.
- Parasitic capacitance between the input and output terminals increase with the operating frequency.

3.12 Transmission Line Impedance Transformers–Guanella and Ruthroff

Transmission line transformers are constructed out of transmission line segments wound into coils in a *bifilar* pattern on a ferrite rod or toroid core. The working principle of a transmission line transformer is the even and odd mode current and voltage waves moving through the two conductors. In the even mode, the current moves in the same direction so that the generated magnetic fields reinforce each other. In the odd mode, the currents propagate in opposite directions along the two conductors, creating equal magnitude but opposite phase magnetic fields that cancel each other, producing a negligible external magnetic field. So a bifilar coil is a large-value inductor for even mode current waves. These even and odd mode currents can be separated by having an *equivalent circuit* consisting of a transmission line segment with two transformers at both ends such that their center taps connected as in Fig. 3.9a. If the inductor L is large, even mode currents are blocked, and only odd mode current flow is allowed.

Several useful signal processing circuits such as a delay, phase inverter and balun (*ba*lanced-*un*balanced) can be constructed with a single bifilar coil. By combining bifilar coils in parallel-series combinations, balun (balanced- unbalanced) and unun (unbalanced-unbalanced) impedance transformers can be constructed, with impedance ratios $1 : n^2$, where n is the number of transmission line segments used.

Figure 3.9b shows a 1:4 Guanella [16] transmission line impedance transformer. *The two transmission line segments are connected so that they are parallel at the low-impedance (source) end and in series at the high-impedance (load) end.* This transmission line impedance transformer connects an unbalanced signal source to a balanced (floating) load. The converse or $n^2 : 1$ (here 4:1) impedance transformer is shown in Fig. 3.9c. For the 1:4 Guanella transmission line impedance transformer,

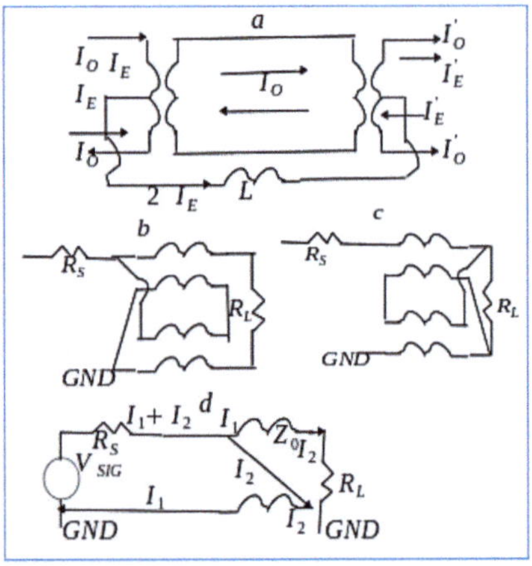

Fig. 3.9 (**a**) Equivalent circuit for a single bifilar coil . (**b, c**) 1:4 and 4:1 Guanella transmission line impedance transformers. (**d**) Ruthroff 1:4 transmission line impedance transformer

each transmission line segment sees only half of the optimum load at any given time. Hence, the optimum characteristic impedance is $Z_0 = \frac{R_{\mathrm{LOAD}}}{4}$. In general, for a $1 : n^2$ Guanella transmission line impedance transformer, the characteristic impedance of each bifilar coil is $\frac{R_{\mathrm{LOAD}}}{(\mathrm{nbf})^2}$, where nbf is the total number of bifilar coils. For a $n^2 : 1$ Guanella transmission line impedance transformer, the characteristic impedance impedance of each bifilar coil is $\mathrm{nbf}^2 R_{\mathrm{LOAD}}$.

The $1 : n^2$ or $n^2 : 1$ Guanella transmission line transformers have a predefined uniform structure, with transmission line terminals on one side (either low impedance or high impedance) connected in series and the other in parallel. Guanella transmission line impedance transformers with fractional or nonuniform termination ratios can be handcrafted, and there is no general formula to design these.

Similar transmission line impedance transformers were proposed and implemented [17] by Ruthroff. *This device adds a direct current (or voltage) with a delayed (by traversing a transmission line) version of the same current (or voltage).* In the Ruthroff 1:4 transmission line impedance transformer (Fig. 3.9d), the direct voltage V_1 is added to the delayed voltage V_2 (after traveling through the transmission line) to obtain the final output voltage across the load. Here, V_1 and V_2 are, respectively, the voltages across the source and load resistances.

When first proposed, both Guanella and Ruthroff transmission line impedance transformers were meant for real sources and loads. However, as demonstrated in Chap. 4, complex loads can be used with both of these, provided their reactances have been neutralized with the Zobel network, as discussed previously.

Exercises

- Extend the analysis for the negative capacitor to a negative inductor.
- Analyze and explain why the series combination of a negative capacitor and a nonnegative capacitor (both with the same magnitude of capacitance value) acts as a short circuit and the parallel combination of a negative capacitor and a nonnegative capacitor (both with the same magnitude of capacitance value) acts as an open circuit.

References

1. Bode HW (1945) Network analysis and feedback amplifier design. Xan Nostrand, New York
2. Fano RM (1948) Theoretical limitations on the broadband matching of arbitrary impedances MTT. Tech. Report No 41 MIT
3. Zobel OJ (1982) Distortion correction in electrical circuits with constant resistance networks. Bell Syst Tech J 7:438
4. Albean D (1995) Zobel networks tame reactive loads EDN p 82
5. Cottee RM, Joines W (1979) Lumped and distributed networks for impedance matching of complex loads. IEEE Trans Circuits Syst 1x-26(5):316–329
6. Matthaei G (1964) Tables of Chebyschev impedance transforming networks of low pass filter form. In: Proceedings of the IEEE

7. Zverev AI Handbook of filter synthesis. - ISBN-10: 0471749427 ISBN-13: 978047174942. Wiley
8. Grebennikov A. RF and microwave power amplifier design - ISBN 0-07-144493-9 Library of Congress – TK7871.58.P6G72 2004 621.395—dc22
9. Carlin HJ, Amstutz P (1981) On optimum broadband matching. IEEE Trans Circuits Syst 28:401–405
10. Carlin HJ, Civalleri PP (1985) On flat gain with frequency-dependent terminations. IEEE Trans Circuits Syst 32:827–839
11. Yarman BS, Fettweis A (1990) Computer-aided double matching via parametric representation of Brune functions. IEEE Trans Circuits Syst 37:212–222
12. Carlin HJ, Yarman BS (1983) The double matching problem: analytic and real frequency solutions. IEEE Trans Circuits Syst 30:15–28
13. Yarman BS (1982) A simplified real frequency technique for broadband matching complex generator to complex loads. RCA Rev 43:529–541
14. Yarman BS (1982) Real frequency broadband matching using linear programming. RCA Rev 43:626–654
15. Youla DC (1964) A new theory of broadband matching. IEEE Trans Circuit Theory 11:30–50
16. Guanella G (1944) New method of impedance matching in radio-frequency circuits. Brown Boveri Rev 31:327–329
17. Ruthroff GL (1959) Some broadband transformers. In: Proceedings of the IRE
18. Linvill JG (June 1953) Transistor negative impedance converters. Proc IRE 41:725–729
19. Perry AK (1973) Broadband antenna systems realized from active circuit conjugate impedance. Naval Postgraduate School, Monterey
20. Kaneko T, Horii Y, Akiyama M (2014) A novel negative impedance converter with an emitter follower circuit for stable non-foster elements. In: EuMC2014, Proceedings, pp 988–991
21. Song KS (2011) Non-Foster impedance matching and loading networks for electrically small antennas. Dissertation of Electrical and Computer Engineering, The Ohio State University
22. Yamanaka S, Utoku Y, Horii Y (2015) A non-Foster negative capacitance generated by parasitic components of operational amplifiers. Korea-Japan Metamater Forum 2015

Chapter 4
Automated Impedance Matching Network Design Process and Design Examples with SPICE Performance Evaluation

4.1 Introduction

The design calculations and equations examined in detail in Chaps. 2 and 3 would not be of much value if they cannot be used to design and implement impedance matching subcircuits for real-world applications. This chapter examines in detail how these complicated design calculations can be automated in the form of C computer language program executables. *Automation guarantees accuracy of results and enables the designer to home in on the optimum solution with a given set of input parameters since the designer does not have to manually do any calculations, thereby allowing for the exploration of the design space.* The only task for the designer is to accurately supply the input parameters to the computer program.

4.2 "L" Lumped Element Impedance Transformer – Automated Design and SPICE Performance Evaluation

The C computer language executable (for both the Linux and Windows operating systems) ***impmatchdirl*** designs and generates the SPICE input netlist for the widely used "L" impedance matching subcircuit (Chap. 2, Sect. 2.1) using a set of user-defined input parameters as operating frequency and load and source resistance and reactance (nonzero). The program checks input validity to ensure that the computed capacitor and inductor values are positive and real valued and finally formats the results as a text SPICE input netlist with a test bench. The program generates the

---

Electronic supplementary material: The online version of this chapter (https://doi.org/10.1007/978-3-319-99001-9_4) contains supplementary material, which is available to authorized users.

A. Banerjee, *Automated Broad and Narrow Band Impedance Matching for RF and Microwave Circuits*, https://doi.org/10.1007/978-3-319-99001-9_4

most common low- and high-pass configuration (of 8 possible) of the "L" impedance transformer. *Both load and source impedances are complex,* i.e., *they have nonzero reactances.* The test bench allows for both SPICE transient and AC (small signal) analyses and enables measurement of RMS (root-mean-squared) input and output power (delivered to the load). Help menus guide the user, and the program can be used in both batch/command line and interactive modes. *Both the source and load impedances are a series combination of a resistor and a capacitor/inductor. Parallel RC or RL pairs may easily be converted into an equivalent series RC or RL counterpart but for analysis purposes only.* The final L impedance matching transformer would of course work with the parallel RC or RL.

Typing **./impmatchdirl** at the command prompt generates the help information as shown. Here, "A" and "B" refer to the two common "L" impedance matching transformer, as shown in Figs. 2.1a, b (Chap. 2):

```
$ ./impmatchdirl
incorrect/insufficient arguments ...
interactive type A L impedance xfrmr
./impmatchdirl i a|A
interactive type B L impedance xfrmr
./impmatchdirl i b|B
batch/command line argument mode ...
./impmatchdirl b|c|B|C a|A|b|B
<load reactance (Ohm) >
<load resistance (Ohm) >
<source reactance (Ohm) >
<source resistance (Ohm) >
<operating frequency (MHz) >
capacitor/inductor reactances/susceptances are negative
sample batch|command line input
./impmatchdirl c A 2 150 -4 50 150
```

For any arbitrary input data set, the program checks if meaningful results can be obtained from the given input set:

```
./impmatchdirl c B -2 75 3 100 250
L xfrmr reactances cannot be computed ...
```

The program is executed from the command line using the sample command line input, **./impmatchdirl c A 2 150 -4 50 150,** which generates the output:

```
B -0.032864 mho 0.003479 mho X 70.856886 Ohm -564.856886 Ohm
negative reactances/susceptances correspond to capacitors/inductors
SPICE netlist file Limpmatchdir.cir
```

The contents of the text SPICE input netlist file are:

```
L IMPEDANCE XFRMR

.PARAMS AMPL=5 FREQ=1.500000e+08
.PARAMS LLIM=7.500000e+07 HLIM=2.250000e+08

.SUBCKT L_XFRMR_A 1 2
* 1 IN
* 2 OUT
C0 1 2 3.230196e-08
L0 2 0 7.521962e-08
.ENDS

** ALTERNATIVE NETLIST

.SUBCKT L_XFRMR_B 1 2
* 1 IN
* 2 OUT
L0 1 2 3.693341e-12
C0 2 0 1.879363e-12
.ENDS

* COMMENT OUT TRANSIENT ANALYSIS
* SECTION TO RUN AC(SMALL SIGNAL)
* ANALYSIS AND VICE-VERSA

**TRANSIENT ANALYSIS
VSIG 1 0 DC 0.0 SIN(0 {AMPL} {FREQ} 0 0 0)
VTST0 3 4 DC 0.0 AC 0.0
VTST1 5 6 DC 0.0 AC 0.0
RSRC 1 2 50.000000
CSRC 2 3 2.653928e-10
XL0 4 5 L_XFRMR_B
RLOAD 6 7 150.000000
LLOAD 7 0 2.123142e-09

.OPTIONS METHOD=GEAR NOPAGE RELTOL=1m MINBREAK=5ps
.IC
.TRAN 15.0ps 10.0us 10.0ps UIC
* INPUT/SOURCE SIDE
.PRINT TRAN V(4) I(VTST0)
* OUTPUT/LOAD SIDE
.PRINT TRAN V(5) I(VTST1)

**AC(SMALL SIGNAL) ANALYSIS
.OPTIONS METHOD=GEAR NOPAGE RELTOL=1m MINBREAK=5ps
VSIG 1 0 DC 0.001 AC {AMPL}
.AC LIN 20000 {LLIM} {HLIM}
.PRINT AC V(5)
.END
```

To measure the RMS (root-mean-squared) power transferred from source to the load at the chosen operating frequency of 150 MHz, the AC analysis section in the above SPICE netlist is commented out, and the measurement is performed in two steps. First, input power is measure, followed by the output power that is delivered to the load. The following two SPICE netlist snippets show this:

```
.OPTIONS METHOD=GEAR NOPAGE RELTOL=1m MINBREAK=5ps
.IC
.TRAN 15.0ps 10.0us 10.0ps UIC
* INPUT/SOURCE SIDE
.PRINT TRAN V(4) I(VTST0)
* OUTPUT/LOAD SIDE
*.PRINT TRAN V(5) I(VTST1)
```

and

```
.OPTIONS METHOD=GEAR NOPAGE RELTOL=1m MINBREAK=5ps
.IC
.TRAN 15.0ps 10.0us 10.0ps UIC
* INPUT/SOURCE SIDE
*.PRINT TRAN V(4) I(VTST0)
* OUTPUT/LOAD SIDE
.PRINT TRAN V(5) I(VTST1)
```

The Ngspice-2.6 simulator is invoked from the command line as:

```
$ ngspice -b Limpmatchdir.cir > out
```

The Ngspice simulator writes the output in tab-separated text format in a simple text file "out." This file contains some metadata/statistics in addition to the simulation output raw data. The metadata is easily edited out, and the supplied C computer language executable utility program **rmscalc** generates the input (to L impedance transformer) and output (delivered to load from L impedance transformer) RMS currents and voltages as:

```
$ ./rmscalc out
2.636008e+00 1.812144e-02 0.000000e+00 666673
```

and

```
$ ./rmscalc out
2.636019e+00 1.757190e-02 0.000000e+00 666675
```

The text output file "out" is generated twice, once for measuring output RMS current/voltage and the second time for generating input RMS current/voltage. The results may be summarized as:

```
RMS input current : 0.01812144 Amperes
RMS input voltage : 2.636008 Volts
RMS output current : 0.0175719 Amperes
RMS output voltage : 2.636019 Volts
```

The ratio of RMS signal output power to RMS signal input power is 0.969678642.

The previous design example for a type "A" L impedance transformer was created by invoking **impmatchdirl** from the command line. Now a type-B L impedance transformer is designed by invoking **impmatchdirl** interactively from the command prompt:

```
./impmatchdirl i b
load reactance(Ohm)
-15
load resistance(Ohm)
75
source reactance(Ohm)
10
source resistance(Ohm)
50
operating frequency(MHz)
450
B -0.013678 mho 0.006131 mho X -0.629961 Ohm -79.370039 Ohm
negative reactances/susceptances correspond to capacitors/inductors
SPICE netlist file Limpmatchdir.cir
```

The contents of the Limpmatchdir.cir are:

```
L IMPEDANCE XFRMR

.PARAMS AMPL=5 FREQ=4.500000e+08
.PARAMS LLIM=2.250000e+08 HLIM=6.750000e+08

.SUBCKT L_XFRMR_A 1 2
* 1 IN
* 2 OUT
C0 1 2 2.587053e-08
.ENDS

** ALTERNATIVE NETLIST
.SUBCKT L_XFRMR_B 1 2
* 1 IN
* 2 OUT
L0 1 2 2.169436e-12
C0 1 0 5.617129e-10
.ENDS

* COMMENT OUT TRANSIENT ANALYSIS
* SECTION TO RUN AC(SMALL SIGNAL)
* ANALYSIS AND VICE-VERSA
```

```
**TRANSIENT ANALYSIS
VSIG 1 0 DC 0.0 SIN(0 {AMPL} {FREQ} 0 0 0)
VTST0 3 4 DC 0.0 AC 0.0
VTST1 5 6 DC 0.0 AC 0.0
RSRC 1 2 50.000000
LSRC 2 3 3.538570e-09
XL0 4 5 L_XFRMR_A
RLOAD 6 7 75.000000
CLOAD 7 0 2.359047e-11

.OPTIONS METHOD=GEAR NOPAGE RELTOL=1m MINBREAK=5ps
.IC
.TRAN 15.0ps 10.0us 10.0ps UIC
* INPUT/SOURCE SIDE
.PRINT TRAN V(4) I(VTST0)
* OUTPUT/LOAD SIDE
.PRINT TRAN V(5) I(VTST1)

**AC(SMALL SIGNAL) ANALYSIS
VSIG 1 0 DC 0.001 AC {AMPL}
.OPTIONS METHOD=GEAR NOPAGE RELTOL=1m MINBREAK=5ps
.AC LIN 20000 {LLIM} {HLIM}
.PRINT AC V(5)
.END
```

The RMS voltage/current across/through the load are:

```
./rmscalc out
2.161524e+00 2.826140e-02 0.000000e+00 666672
```

Also, the RMS voltage/current at the end of the source impedance (right before the L impedance transformer) are:

```
./rmscalc out
2.161600e+00 2.826140e-02 0.000000e+00 666672
```

where "out" is an intermediate text data file generated by directing the output of the Ngspice simulator to a text file. It is generated twice, one for measuring output RMS current/voltage and the second time for generating input EMS current/voltage. Summarizing

```
RMS input current 0.02826140 Amperes
RMS input voltage 2.161600 Volts
Total number of data sample values 666672
RMS output current 0.02826140 Amperes
RMS output voltage 2.161524 Volts
Total number of data sample values 666672
```

Clearly, almost the entire input signal power is delivered to the load.

There are two drawbacks to using lumped element circuits:

- RF and microwave frequencies (100 s of MHz to 10 s of GHz) exceed the cut-off frequencies of some discrete capacitors and inductors.
- Fabricating very low-value inductors (e.g., 2.5e-12 Henry) is very difficult because the fabrication technology is the same as that used for making printed circuit boards and results in lossy inductors.

Finally, if the ratio of RMS output to RMS input power is below the required threshold, the values of the L transformer's capacitor and inductor can be modified as required to satisfy the threshold/target. The source and load impedances **cannot be modified** in any way.

4.3 Single Transmission Line Segment Impedance Transformer – Automated Design and SPICE Performance Evaluation

The C computer language executable **impmatchtxline** calculates the characteristic impedance and electrical and physical lengths of a single transmission line segment that can match any arbitrary complex source–load impedance pair, along with the text SPICE input netlist and test bench for both ideal and lossy transmission line models. It determines if the input complex source–load impedance pair can be matched or not and, if yes, computes the corresponding values for the characteristic impedance and electrical and physical length. The output is formatted as text SPICE input netlist, with both the default lossless and user-supplied lossy transmission line models. The associated SPICE test bench allows for RMS power transfer (source to load) measurement with both the lossless and lossy transmission line models. The same test bench also allows measurement of reflection coefficient, insertion loss, and standing wave ratio (SWR) for both the lossless and lossy transmission line models. **impmatchtxline** can be executed in both batch/command line and interactive modes. Typing **./impmatchtxline** at the command prompt generates the following help information:

```
$ ./impmatchtxline
Incorrect/insufficient arguments
for interactive mode
./impmatchtxline i
for batch/command line argument mode
./impmatchtxline b|c|B|C <frequency(MHz)>
<load resistance(Ohm)>
<load reactance(Ohm)>
<source resistance(Ohm)>
<source reactance(Ohm)>
sample batch|command line input
./impmatchtxline B 200 50 10 65 -5
```

Using the sample command line input generates the following output:

```
./impmatchtxline B 200 50 10 65 -5
Computed transmission line parameters
characteristic impedance 53.851648 Ohm
physical length 0.174709 m
SPICE netlist file : Txlineimm.cir
```

So the characteristic impedance of the required transmission line segment is approximately 54.0 ohms, and the physical length is 0.17 m or 17.0 cm. The contents of the generated text SPICE input netlist are:

```
SINGLE TX. LINE IMPEDANCE MATCHING

.PARAMS FREQ=2.000000e+08 TD=5.823626e-10 Z0=53.851648
AMPL=10.000000

.PARAMS RL=50.000000 XL=7.961783e-09
.PARAMS RS=65.000000 XS=1.592357e-10
.PARAMS LLIM=1.000000e+08 HLIM=3.000000e+08
** LOSSY TRANSMISSION LINE MODEC-- CAN BE MOFIFIED
.PARAMS R=0.0175 L=120.0E-9 G=0.0
.PARAMS C=120.0E-12 LEN=0.174709

.MODEL LTXL LTRA R={R} L={L}
+ G={G} C={C} LEN={LEN}
* COMMENT OUT TRANSIENT ANALYSIS
* SECTION TO RUN AC/SMALL SIGNAL
* ANALYSIS AND VICE-VERSA
* TRANSIENT ANALYSIS IDEAL
VSIG 1 0 DC 0.001 SIN(0 {AMPL} {FREQ} 0 0 0)
VTST0 3 4 DC 0.0 AC 0.0
VTST1 5 6 DC 0.0 AC 0.0
RS 1 2 {RS}
CSRC 2 3 1.592357e-10
T0 4 0 5 0 Z0={Z0} TD={TD}
LLOAD 6 7 7.961783e-09
RL 7 0 {RL}
* TRANSIENT ANALYSIS LOSSY
VSIG 1 0 DC 0.001 SIN(0 {AMPL} {FREQ} 0 0 0)
VTST0 3 4 DC 0.0 AC 0.0
VTST1 5 6 DC 0.0 AC 0.0
RS 1 2 {RS}
CSRC 2 3 1.592357e-10
O1 4 0 5 0 LTXL
LLOAD 6 7 7.961783e-09
RL 7 0 {RL}
** TRANSIENT ANALYSIS IDEAL/LOSSY
.OPTIONS METHOD=GEAR NOPAGE RELTOL=1m MINBREAK=5ps
.IC
.TRAN 20.0ps 300.0ns 10.0ps UIC
```

```
* INPUT/SOURCE SIDE
.PRINT TRAN V(4) I(VTST0)
* OUTPUT/LOAD SIDE
.PRINT TRAN V(5) I(VTST1)

** AC/SMALL SIGNAL ANALYSIS
VSIG 1 0 DC 0.001 AC {AMPL}
.OPTIONS METHOD=GEAR NOPAGE RELTOL=1m MINBREAK=5ps
.AC LIN 20000 {LLIM} {HLIM}
.PRINT AC V(2)
.END
```

The results of the ideal/lossless transmission line SPICE transient analysis to measure RMS power transfer from source to load are:

```
RMS input current : 0.05498559 Amperes
RMS input voltage : 3.579869 Volts
Total number of sample data values : 15820

RMS output current : 0.06216719 Amperes
RMS output voltage : 3.183085
Total number of sample data values : 15820
```

The ideal-transmission-line-based impedance transformer delivers almost 100% of the signal power from source to load. The reflection coefficient, the insertion loss, and the standing wave ratio (SWR) for the ideal-transmission-line-based design are shown in Figs. 4.1, 4.2, and 4.3.

The RMS signal power transferred using the unmodified lossy transmission line segment is summarized below:

```
RMS input current : 0.07106488 Amperes:
RMS input voltage : 2.569631 Volts
Total number of sample data values : 15009
RMS output current : 0.05685381 Amperes
RMS output voltage : 2.899925 Volts
Total number of sample data values : 15009
```

Therefore, the ratio of the output RMS power (delivered to load from source) to the input RMS power is 0.902860285. If this ratio is below the preset threshold, it can be increased (thereby increasing the power delivered from source to load) by judiciously (through experimentation) modifying lossy transmission line parameters (capacitance/length, conductance/length, inductance/length, and resistance/length) of the SPICE netlist file as:

```
** LOSSY TRANSMISSION LINE MODEL - CAN BE MODIFIED

.PARAMS R=0.015 L=1.0E-13 G=0.0
.PARAMS C=1.0E-12 LEN=0.1
.MODEL LTXL LTRA R={R} L={L}
+ G={G} C={C} LEN={LEN}
```

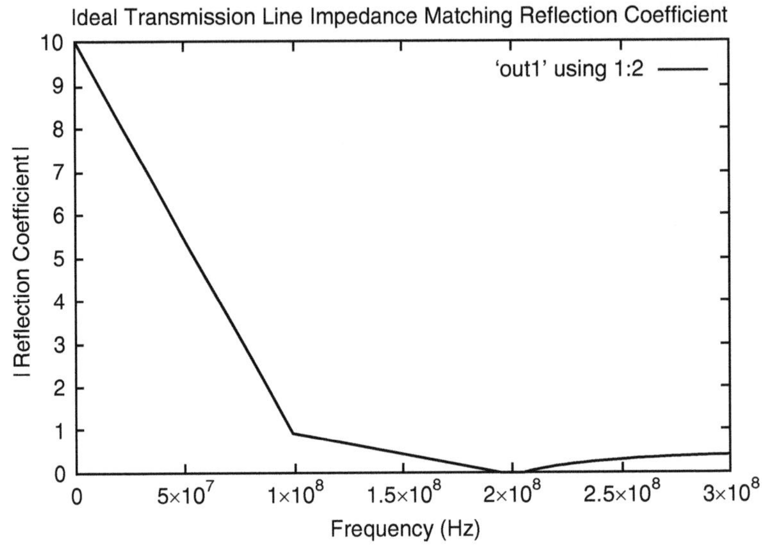

Fig. 4.1 Ideal transmission line impedance transformer reflection coefficient

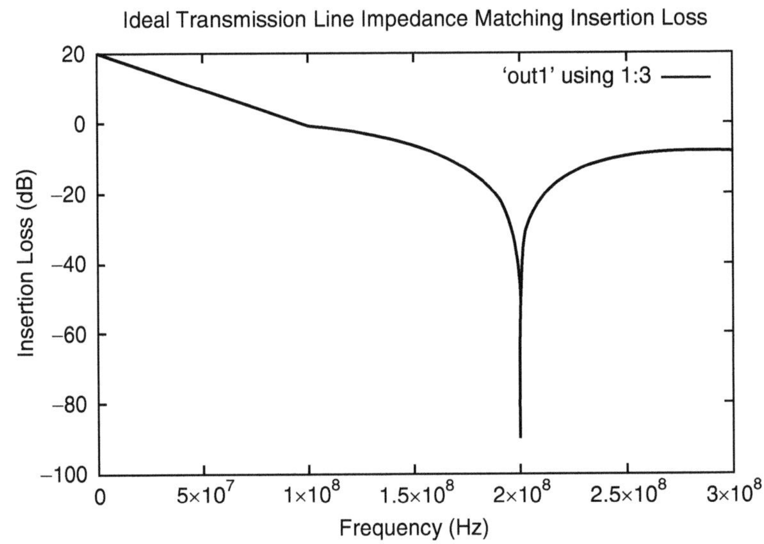

Fig. 4.2 Ideal transmission line impedance transformer insertion loss

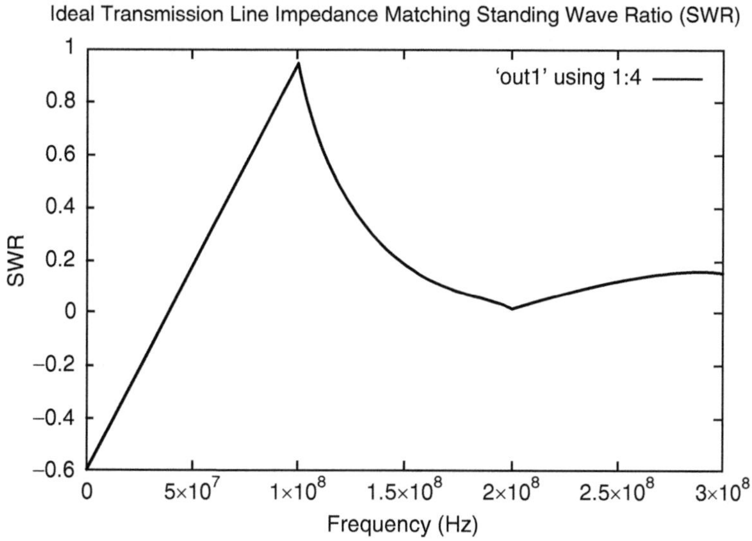

Fig. 4.3 Ideal transmission line impedance transformer standing wave ratio

Then the results of the transient analysis to measure RMS signal power transferred to the load are:

```
RMS input current : 0.04854972 Amperes
RMS input voltage : 2.945541 Volts
Total number of sample data values : 31154
RMS output current :0.04829070 Amperes
RMS output voltage : 2.945468 Volts
Total number of sample data values : 31154
```

The modified lossy-transmission-line-based impedance transformer delivers almost 100% of the signal power from source to load.

4.4 Single/Double Closed-/Open-Circuit Transmission Line Stub Impedance Transformer – Automated Design and SPICE Performance Analysis

The C computer language executable **impmatchtxlstub** does the calculations to design and evaluate the performance characteristics of single/double closed-/open-circuit parallel/series transmission line stub impedance matching subcircuit. It can be executed in both batch/command line and interactive mode, and outputs the text SPICE netlist and test bench. It uses both the SPICE built-in lossless and user-defined lossy transmission line models. The SPICE netlist enables the measurement of RMS signal power transferred from the source to the load, as well as the reflection coefficient, insertion loss, and SWR for both the lossy- and lossless-transmission-line-based single/double closed/open stub impedance transformer. The load is complex. Typing **./impmatchtxlstub** at the command prompt generates the help information:

```
incorrect/insufficient arguments ...
interactive single parallel closed/open circuit stub
./impmatchtxlstub i p
interactive single series closed/open circuit stub
./impmatchtxlstub i s
double open circuit stubs
./impmatchtxlstub i d
batch/command line argument mode
./impmatchtxlstub b|B|c|C p|s|d
<characteristic impedance(ohm)>
<operating frequency(MHz)>
<load reactance(Ohm)>
<load resistance(Ohm)>
<stub separation(fraction of wavelength)>
STUB SEPARATION IS NON-ZERO ONLY FOR DOUBLE STUB CONFIGURATION
sample batch/command line input for single parallel stub configuration
./impmatchtxlstub b p 50 250 -10 75 0
```

Using the sample batch/command line input for a single, parallel closed-/open-circuited stub impedance transformer generates the output:

```
$ ./impmatchtxlstub b p 50 250 -10 75 0
general case parallel stub
tx. line segment lengths(multiples of wavelength)
d0 0.118354 d1 0.333768
stub lengths(multiples of wavelength)
open circuit 0.434036 short circuit 0.315837
SPICE input netlist file : Txlsgldblstub.cir
```

The SPICE netlist allows the measurement of RMS signal power delivered from source to load, as well as reflection coefficient and insertion loss for each of the design cases listed below:

Parallel single open-circuit stub, ideal transmission line

```
RMS input current : 0.07243459 Amperes
RMS input voltage : 3.551240 Volts
Total number of sample data values : 7262
RMS output current : 0.05764008 Amperes
RMS output voltage : 4.355040 Volts
Total number of sample data values : 7262
```

In this case, 0.975867137 of the RMS input power is transferred to the load.

Parallel single closed-circuit stub, ideal transmission line

```
RMS input current : 0.07053968 Amperes
RMS input voltage : 3.465757 Volts
Total number of sample data values : 7279
RMS output current : 0.05745436 Amperes
RMS output voltage : 4.336535 Volts
Total number of sample data values : 7278
```

Therefore, almost all RMS input signal power is transferred to the load.

Parallel single open-circuit stub, lossy transmission line

```
RMS input current : 0.08067865 Amperes
RMS input voltage : 3.295867 Volts
Total number of sample data values : 6679
RMS output current : 0.05420635
RMS output voltage : 4.107264 Volts
Total number of sample data values : 6678
```

As expected, the lossy transmission line increases input signal power loss, and 0.83728726 of the RMS input signal power is delivered to the load.

Parallel single closed-circuit stub, lossy transmission line

```
RMS input signal current : 0.1117020 Amperes
RMS input voltage : 1.65117 Volts
Total number of sample data values : 6677
RMS output current : 0.04577198 Amperes
RMS output voltage : 3.463911 Volts
Total number of sample data values : 6677
```

Once again, the lossy transmission line increases loss of input signal power so that only 0.859634201 of the RMS input signal power is delivered to the load.

The reflection coefficients and insertion losses for each of the four cases are shown next (Figs. 4.4, 4.5, 4.6, 4.7, 4.8, 4.9, 4.10, and 4.11).

Identical analysis may be done for the double closed/open or mixed mode ideal/lossy transmission line stub model. The user-defined lossy transmission line parameters as capacitance/length, inductance/length, and resistance/length can be modified to make the lossy transmission line stub impedance transformer (single/double stub, closed/open circuit) performance characteristics match those of the ideal transmission line transformer case.

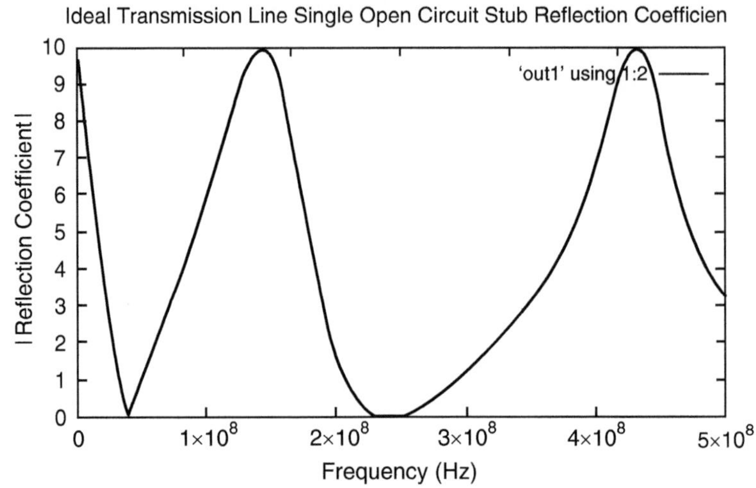

Fig. 4.4 Ideal transmission line single open-circuit stub reflection coefficient

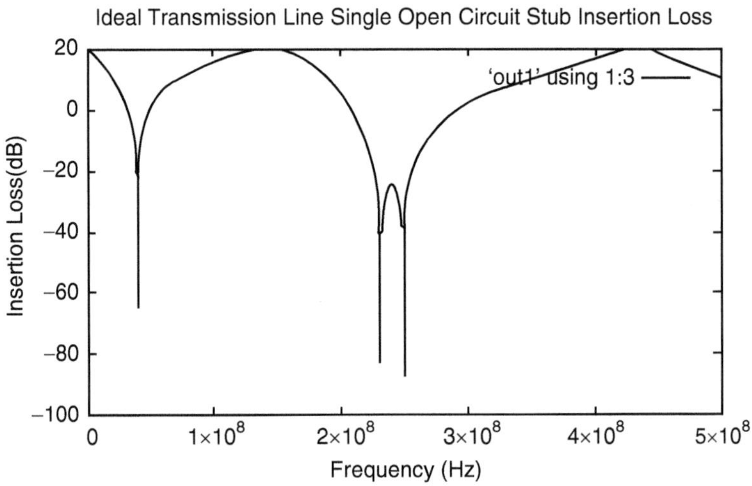

Fig. 4.5 Ideal transmission line single open-circuit stub insertion loss

4.5 "Convert" Complex Load to Resistance – Parallel/Series Capacitor or Inductor Load Reactance – Automated Design and SPICE Performance Analysis

The key distinguishing feature about broadband impedance matching is that the reactances of both a capacitor and an inductor vary with the signal frequency, and so the voltage across and current through a complex load vary over a broad frequency range. Consequently, the gain and therefore the signal power delivered

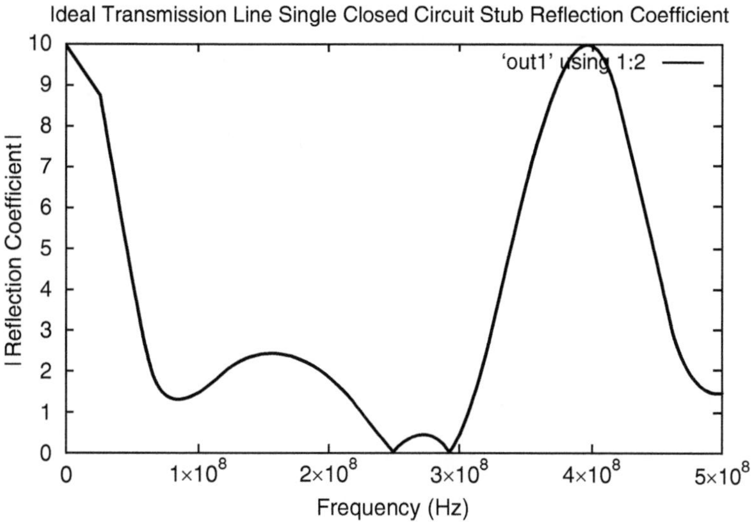

Fig. 4.6 Ideal transmission line single closed-circuit stub reflection coefficients

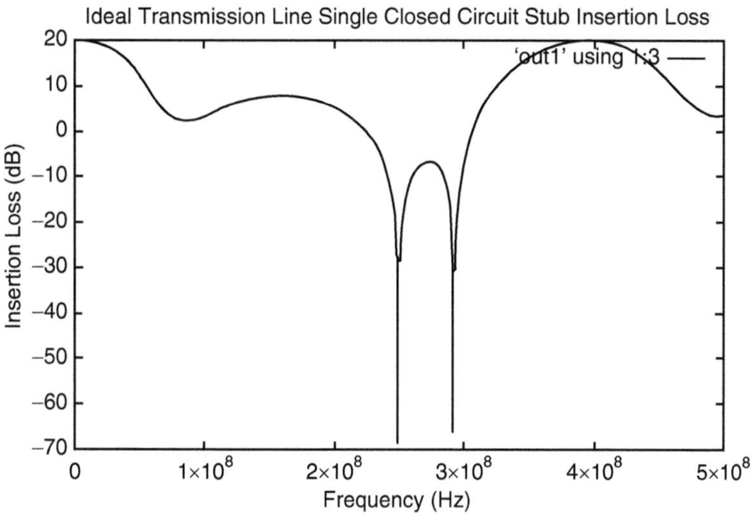

Fig. 4.7 Ideal transmission line single closed-circuit stub insertion loss

to the load would vary over the same broad frequency range. ***So the reactive component of a complex load (parallel or series RL or RC) must be neutralized so that the complex impedance appears as a resistor to the signal source. With the frequency dependency of the complex load nullified, the gain would be a constant (within applicable tolerances).*** The Zobel network allows the designer to achieve this goal.

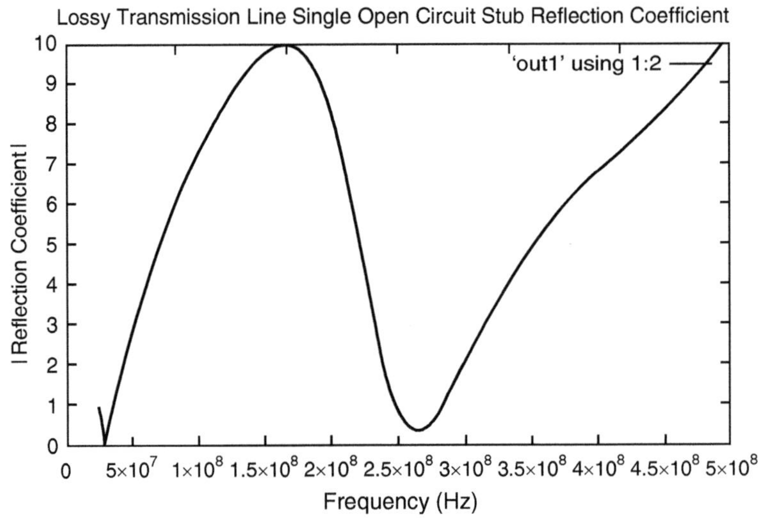

Fig. 4.8 Lossy transmission line single open-circuit stub reflection coefficient

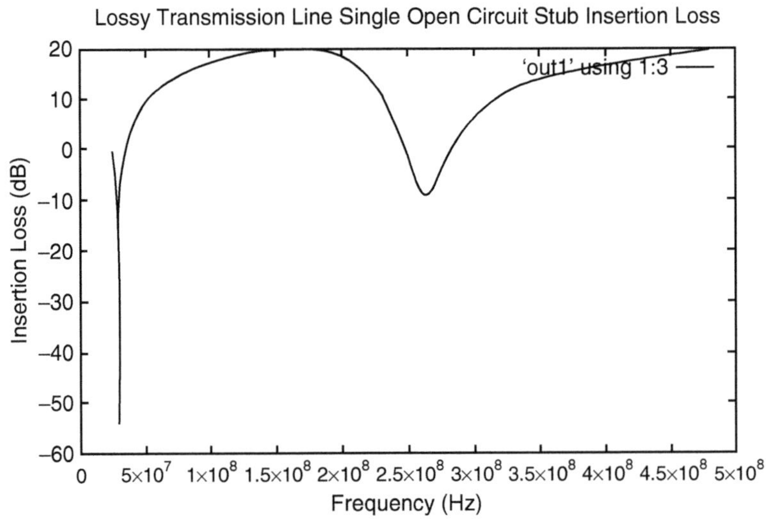

Fig. 4.9 Lossy transmission line single open-circuit stub insertion loss

The C computer language program **impmatchbbpszobel** computes the component values for the Zobel network for a given parallel or series complex impedance. *As the Zobel network was initially conceived of to tackle series complex impedances only,* ***impmatchbbpszobel*** *first checks if the input complex impedance is series or parallel RC or RL, and then if it is a parallel RC or RL, it converts it to the equivalent series RC or RL at the band center frequency. Then it uses the new equivalent series RC or RL component values to compute the corresponding Zobel network.* As the

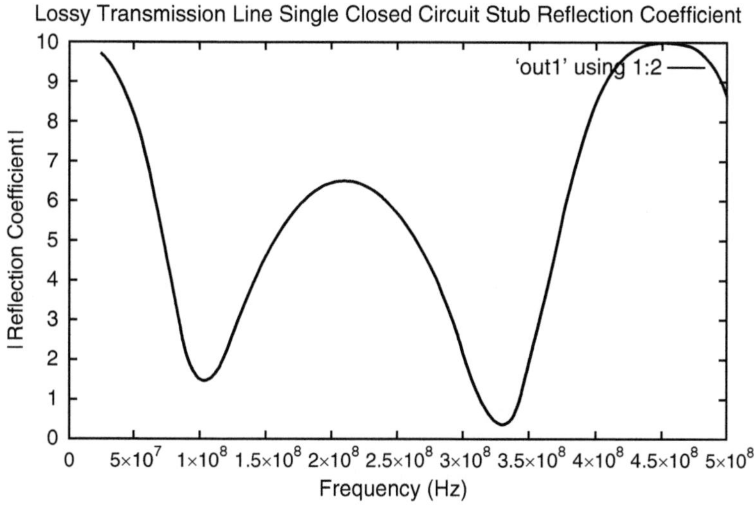

Fig. 4.10 Lossy transmission line single closed-circuit stub reflection coefficient

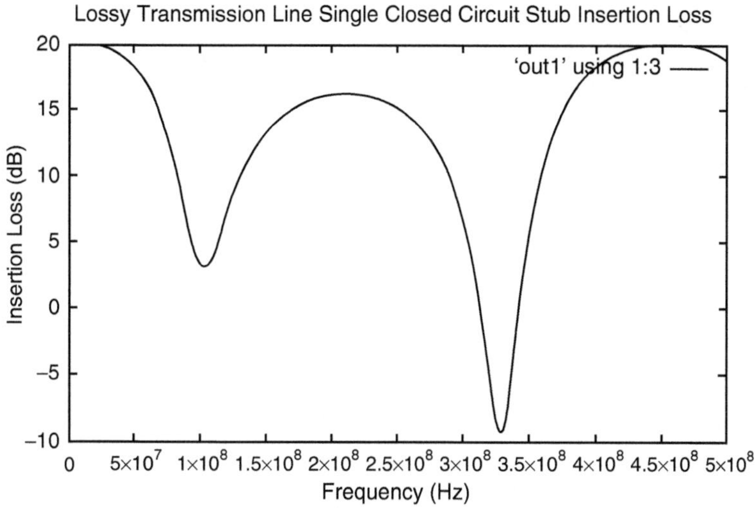

Fig. 4.11 Lossy transmission line single closed-circuit stub insertion loss

Zobel network designer cannot modify the configuration (parallel or series RC or RL) of a given load impedance, for parallel load impedances, a hybrid Zobel network is used. This will be examined in detail subsequently. The final results are formatted as a SPICE text input netlist that can be used with any SPICE simulator for performance analysis. **Impmatchbbpszobel** can be executed in both batch/command line and interactive modes. Typing **./impmatchbbpszobel** at the command prompt generates the help information:

```
$ ./impmatchbbpszobel
incorrect/insufficient arguments
interactive mode
./impmatchbbpszobel i
batch/command line argument mode
./impmatchbbpszobel b|B|c|C
<frequency band low limit(MHz)>
<frequency band high limit(MHz)>
<load reactance (picoFarad|nanoHenry)>
<load resistance(Ohm)>
<load reactance type capacitor/inductor (1|2)>
<load configuration - series/parallel RC/RL (1|2)>
sample command line input parallel RC load
./impmatchbbpszobel b 500 800 15 50 1 2
sample command line input parallel RL load
./impmatchbbpszobel b 250 850 20 75 2 2
```

Using the batch/command line input for the sample parallel RC impedance generates:

```
$ ./impmatchbbpszobel b 500 800 15 50 1 2
parallel RC | RL load
SPICE netlist file : serparzobel.cir
```

The contents of the SPICE netlist file serparzobel.cir are:

```
PARALLEL RC - RL Zobel network
.PARAMS FLLIM=5.000000e+08 FHLIM=8.000000e+08 AMPL=10

.SUBCKT PARZOBEL 1 2
** 1 IN
** 2 OUT
R0 2 3 0.000071
C0 3 0 1.499998e-11
R1 1 2 1
R2 1 4 0.000071
LCOMP 4 0 2.976516e-03
.ENDS

VSIG 1 0 DC 0.001 AC {AMPL}
RS 1 2 0.000071
XZ0 2 3 PARZOBEL

.OPTIONS METHOD=GEAR NOPAGE RELTOL=1m MINBREAK=5ps
.AC LIN 50000 {FLLIM} {FHLIM}
.PRINT AC V(3)
.END
```

The frequency and phase responses of the above Zobel network are shown in Figs. 4.12 and 4.13, respectively.

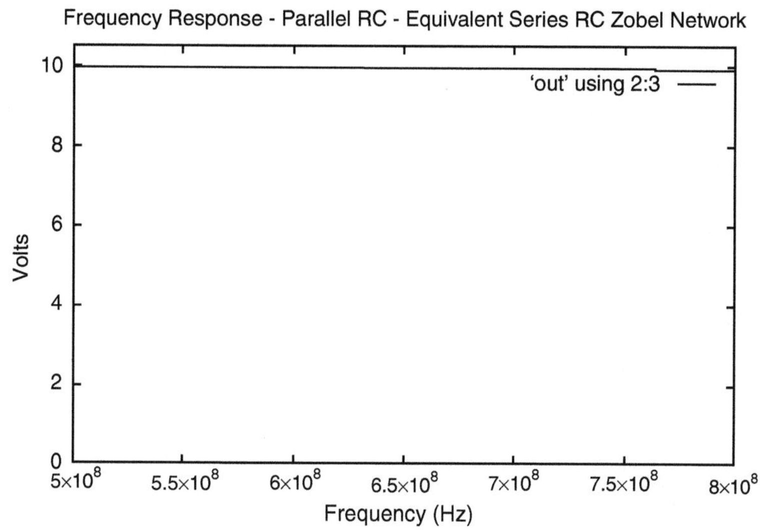

Fig. 4.12 Frequency response of Zobel network of equivalent series RC complex load, transformed from original parallel RC impedance

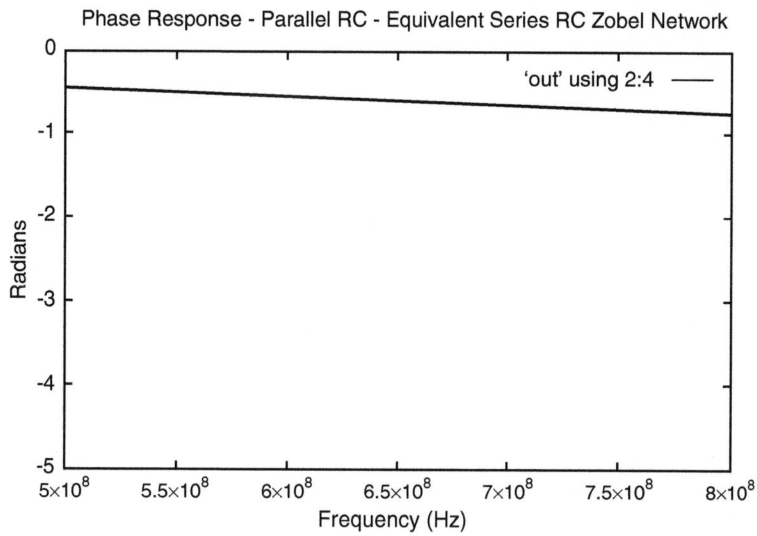

Fig. 4.13 Phase response of Zobel network of equivalent series RC complex load, transformed from original parallel RC impedance

Similarly, using the batch/command line sample input for parallel RL impedance, one gets:

```
$ ./impmatchbbpszobel b 250 850 20 75 2 2
parallel RC | RL load
SPICE netlist file : serparzobel.cir
```

and the contents of the SPICE netlist are listed below:

```
PARALLEL RC - RL Zobel network
.PARAMS FLLIM=2.500000e+08 FHLIM=8.500000e+08 AMPL=10

.SUBCKT PARZOBEL 1 2
** 1 IN
** 2 OUT
R0 2 3 28.006283
L0 3 0 1.253166e-08
R1 1 2 1
R2 1 4 28.006283
CCOMP 4 0 1.597709e-11
.ENDS

VSIG 1 0 DC 0.001 AC {AMPL}
RS 1 2 28.006283
XZ0 2 3 PARZOBEL

.OPTIONS METHOD=GEAR NOPAGE RELTOL=1m MINBREAK=5ps
.AC LIN 50000 {FLLIM} {FHLIM}
.PRINT AC V(3)
.END
```

Any available SPICE simulator can be used to generate the corresponding frequency and phase responses for this netlist. This is left as an exercise for the reader. The sequence of calculation steps to implement the Zobel network for a real-world parallel RC or RL impedance is summarized in Fig. 4.14. For both the previous two Zobel network design examples, **impmatchbbpszobel** was invoked in the batch/command line argument mode. So for the next design example, it is executed/invoked interactively to create the Zobel network for a series RC impedance. The interactive data input is shown next, and the frequency response is shown in Fig. 4.15.

```
$./impmatchbbpszobel I
frequency band low limit(MHz)
300
frequency band high limit(MHz)
750
load reactance (pF/nH)
15
load resistance(Ohm)
60
```

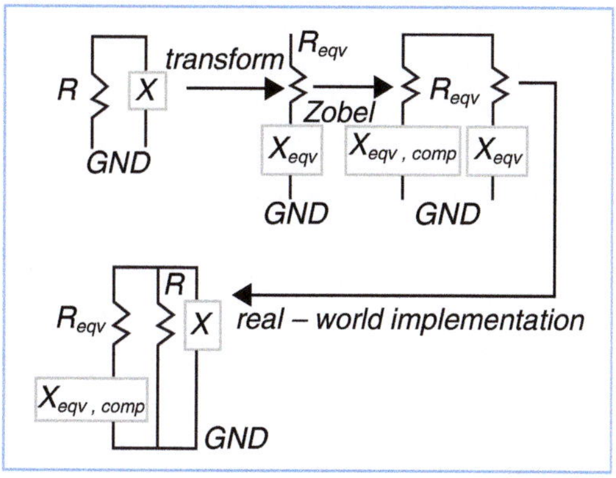

Fig. 4.14 Calculation and transformation steps to implement the Zobel network for a parallel RC or RL impedance

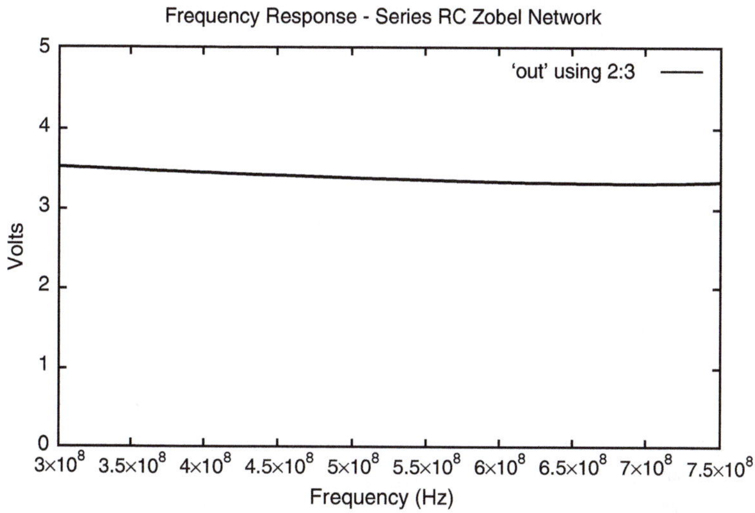

Fig. 4.15 Frequency response series RC impedance converted to Zobel network

```
load reactance type capacitor/inductor (1|2)
1
load configuration - series/parallel RC/RL (1|2)
1
series RC | RL load
SPICE netlist file : serparzobel.cir
```

To reinforce the concepts underlying the sequence of steps in Fig. 4.14, the C computer language executable **impmatchpsrcrl** generates two SPICE netlists for a given parallel RC or RL input:

- Straightforward parallel RC|RL transformation to equivalent series RC|RL, followed by Zobel network for the equivalent series RC with SPICE netlist and test bench

- Straightforward parallel RC|RL transformation to equivalent series RC|RL, followed by the hybrid Zobel network for the equivalent series RC with SPICE netlist and test bench. The *hybrid Zobel network has its equivalent series RC|RL leg replaced by the original parallel RC|RL.*

So using the command line input **$ ./impmatchpsrcrl b 400 15 50 1** generates the Zobel and hybrid Zobel networks as *it must be noted that the SPICE netlist for the hybrid Zobel network can be constructed very easily by inspection of the SPICE netlist of the Zobel network for the equivalent series RC|RL corresponding to a parallel RC|RL, as generated by* ***impmatchbbpszobel***:

```
.SUBCKT SRCZOBEL 1 2
** 1 IN
** 2 OUT
CSER 3 0 1.922600e-11
LCOMPSER 4 0 1.591724e-13
RSER 2 3 10.990325
RSERCOMP 1 4 10.990325
R0 1 2 1.000000
.ENDS

.SUBCKT PRCHYBRIDZOBEL 1 2
** 1 In
** 2 OUT
CPAR 2 0 1.500000e-11
LCOMPSER 4 0 1.591724e-13
RPAR 2 0 50.000000
RSERCOMP 1 4 10.990325
R0 1 2 1.000000
.ENDS
```

The frequency response lots for the equivalent series RC Zobel network and the hybrid equivalent series RC Zobel network are shown in Figs. 4.16 and 4.17, respectively.

Clearly, the frequency responses of the equivalent RC Zobel and hybrid equivalent RC Zobel networks (Figs. 4.16 and 4.17) appear very similar, with the output voltage varying within the same range, over the same frequency band, thereby reinforcing the initial premise that ***for calculation of the Zobel network components only, the parallel RC|RL can be replaced by its equivalent series RC|RL counterpart, but at the actual implementation stage, only the compensating reactance and compensating resistance of the Zobel network can be used with the original parallel RC|RL.*** **This is a key result because in real-world design scenarios, the designer cannot modify the load impedance configuration (parallel|series RC|RL) at all, and the one and only one way (the alternative is to use an active grounded negative impedance converter) to neutralize the reactance of a parallel RC|RL load is to use the hybrid Zobel network approach, as explained above.**

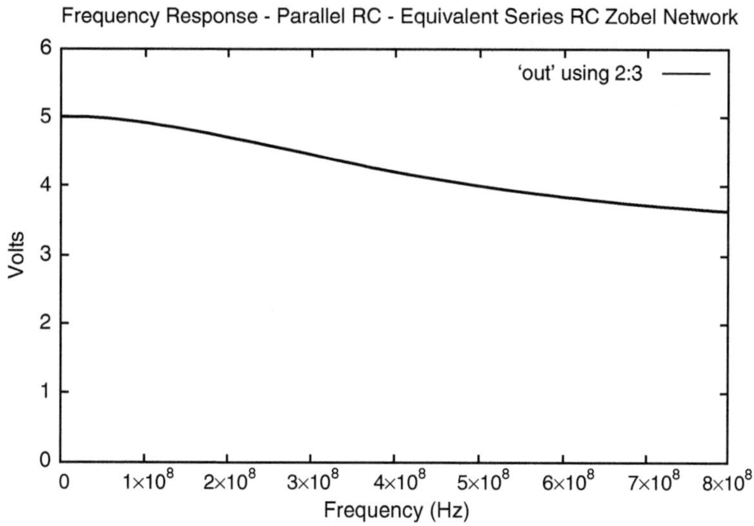

Fig. 4.16 Frequency response for parallel RC to equivalent series RC transformation and Zobel network

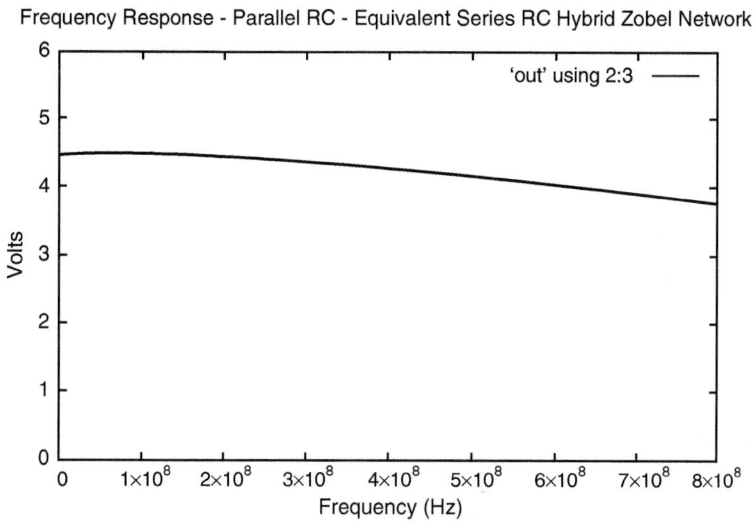

Fig. 4.17 Frequency response for parallel RC to equivalent series RC transformation and hybrid Zobel network. The hybrid Zobel network has the equivalent series leg replaced by the original parallel RC

The Zobel network is the most effective subcircuit to implement broadband impedance matching since by "converting" a parallel RC|RL load to its equivalent series RC|RL load, it enables the construction of the Zobel network that allows the designer to circumvent the bandwidth-gain limitations imposed by the Bode–Fano inequalities. Also, the Zobel network was originally conceived of for series RC|RL loads, and so the same advantage of not having to tackle bandwidth-gain limits of Bode–Fano inequalities applies in this case as well. **impmatchpsrcrl** can be executed in batch/command line argument mode and has help menus to aid the user.

4.6 Active Impedance Matching with Bi-junction Transistors in Feedback Loop and Shorted Transmission Line Segments to Neutralize Reactances

The C computer language executable **impmatchnegimpconv** uses designer-supplied SPICE RF and microwave bi-junction transistor models to generate the corresponding text SPICE netlists for both the floating and grounded negative impedance converters for a given capacitor or inductor (the reactive part of a series or parallel RC or RL pair), along with the respective test benches. *As the performance characteristics of any negative impedance converter is controlled by the properties of the bi-junction transistor used in it, the designer has to ensure that the operating frequency range of the transistor matches the frequency range over which the negative impedance converter has to operate.* **impmatchnegimpconv** can be executed in both batch/command line argument mode and interactive mode. Initially, negative impedance converters were conceived of to neutralize the reactance of a series RC or RL load, e.g., an antenna. However, **impmatchnegimpconv** allows the designer to analyze the case of a parallel RC or RL load by first converting it to the equivalent series RC or RL impedance and then generating the negative impedance converter for the equivalent series RC or RL. Typing **impmatchnegimpconv** at the command prompt generates the help message:

```
$ ./impmatchnegimpconv
incorrect/insufficient arguments
interactive mode
./impmatchnegimpconv i
batch/command line argument mode
./impmatchnegimpconv b|B|c|C
<frequency band low limit(MHz)>
<frequency band high limit(MHz)>
<load reactance capacitor|inductor (pF|nH)>
<load resistance(Ohm)>
<load reactance type capacitor|inductor (1|2)>
<load configuration series|parallel RC|RL (1|2)>
sample batch/command line input for series RC load
./impmatchnegimpconv b 100 500 15 50 1 1
sample batch/command line input for parallel RL load
./impmatchnegimpconv C 100 500 15 50 2 2
```

In the subsequent discussion, a series RC or RL load is used with a floating negative impedance converter, and a grounded negative impedance converter is always used with a shunt RC or RL load. Using the command line input to design a negative capacitance converter for a series EC load, **impmatchnegimpconv** generates:

```
$ ./impmatchnegimpconv b 150 750 10 50 1 1
SPICE netlist text file : negimpcapind.cir
```

where the contents of the text SPICE input are:

```
NEGATIVE CAPACITOR

** SPICE RF|MICROWAVE TRANSISTOR
** MODEL FILE MUST BE IN SAME DIRECTORY
** PATH AS NETLIST

.INCLUDE hiperftransistormodels

.PARAMS FLLIM=1.500000e+08 FHLIM=7.500000e+08 LDREACT=1.000000e-11
.PARAMS LDRESIS=5.000000e+01 MMR=1.000000e+08 AMPL=10

** GROUNDED NEGATIVE CAPACITOR
.SUBCKT NEGCAPGND 1 2
** 1 IN
** 2 OUT - GROUNDED
C0 3 2 {LDREACT}
R0 4 2 50.0
R1 5 2 50.0
XQ0 3 4 1 BFR92A
XQ1 4 3 5 BFR92A
.ENDS

** FLOATING NEGATIVE CAPACITOR

.SUBCKT NEGCAPFLOAT 1 2
** 1 IN
**2 OUT
C0 3 4 {LDREACT}
XQ0 3 4 1 BFR92A
XQ1 4 3 2 BFR92A
.ENDS

** COMMENT OUT GOUNDED CAPACITOR TEST BENCH
** TO EXECUTE FLOATING CAPACITOR TEST BENCH
** AND VICE-VERSA

** GROUMDED NEGATIVE CAPACITOR TEST BENCH
C0 2 3 {LDREACT}
R0 1 2 {LDRESIS}
```

```
R1 3 4 {LDRESIS}
XNCG 4 0 NEGCAPGND

** FLOATING NEGATIVE CAPACITOR TEST BENCH
C0 2 3 {LDREACT}
R0 1 2 {LDRESIS}
R1 4 5 {LDRESIS}
R2 5 0 {MMR}
XNCF 3 4 NEGCAPFLOAT

VSIG 1 0 DC 0.0001 AC {AMPL}
.OPTIONS METHOF=GEAR NOPAGE RELTOL=1m MINBREAK=5ps
.AC LIN 50000 {FLLIM} {FHLIM}
.PRINT AC V(4)
.END
```

The generated frequency response is shown in Fig. 4.18.

It is evident from Fig. 4.18 that the floating negative capacitor for the series RC load impedance provides perfect impedance matching over the entire operating frequency range. ***A floating negative capacitor in series with a series floating RC load represents a monopole antenna, with the antenna insulated/isolated from a ground plane—SPICE represents this by having a very large resistance to ground, effectively an floating/open circuit. A dipole antenna is simulated as two isolated monopole antennas.***

Now **impmatchnegimpconv** is used to design a negative impedance converter for a shunt RL load, using the command line arguments:

```
$ ./impmatchnegimpconv C 200 800 15 50 2 2
SPICE netlist text file : negimpcapind.cir
```

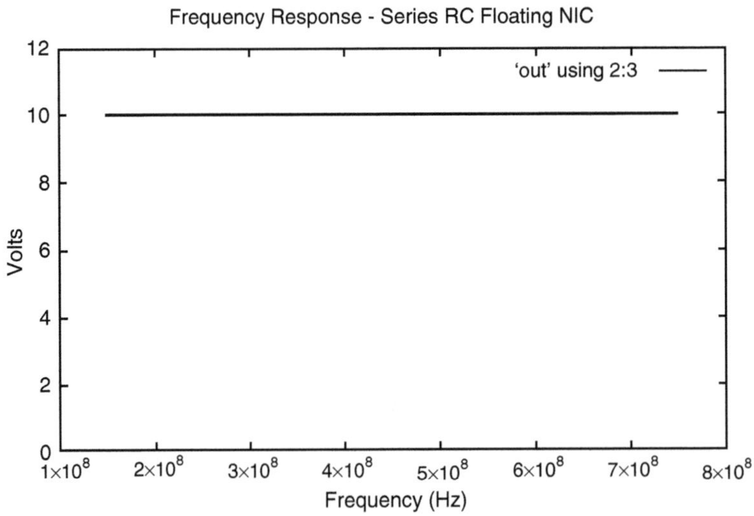

Fig. 4.18 Frequency response of floating negative impedance converter for series RC load

where the contents of the generated text SPICE netlist are:

```
GROUNDED NEGATIVE INDUCTOR SHUNT RL INDUCTOR

** SPICE RF|MICROWAVE TRANSISTOR
** MODEL FILE MUST BE IN SAME DIRECTORY
** PATH AS NETLIST

.INCLUDE hiperftransistormodels

.PARAMS FLLIM=2.000000e+08 FHLIM=8.000000e+08 LDREACT=1.500000e-08
.PARAMS LDRESIS=5.000000e+01 MMR=1.000000e+08 UL=1.0 AMPL=10

** GROUNDED NEGATIVE INDUCTOR
.SUBCKT NEGINDGND 1 2
** 1 IN
** 2 OUT - GROUNDED
C0 3 2 {LDREACT}
R0 4 2 19.0
R1 5 2 19.0
XQ0 3 1 4 BFR92A
XQ1 1 3 5 BFR92A
.ENDS

.SUBCKT NEGINDFLOAT 1 2
** 1 IN
**2 OUT
L0 3 4 {LDREACT}
XQ0 3 4 1 BFR92A
XQ1 4 3 2 BFR92A
.ENDS

** GROUMDED NEGATIVE INDUCTOR TEST BENCH
L0 2 0 {100.9*LDREACT}
R0 1 2 {LDRESIS}
R1 2 3 {UL}
R2 3 0 {LDRESIS}
R3 3 4 {UL}
XNIG 4 0 NEGINDGND

VSIG 1 0 DC 0.0001 AC {AMPL}
.OPTIONS METHOF=GEAR NOPAGE RELTOL=1m MINBREAK=5ps
.AC LIN 50000 {FLLIM} {FHLIM}
.PRINT AC V(4)
.END
```

The corresponding frequency response for the grounded negative inductor for the shunt RL load is shown in Fig. 4.19.

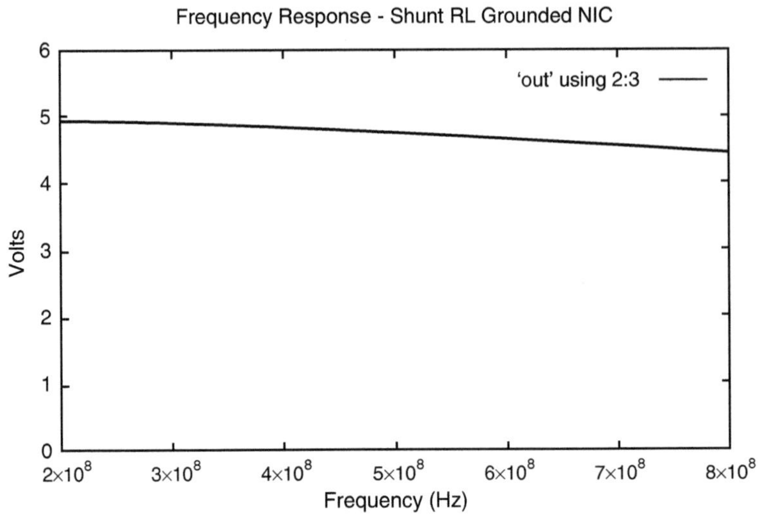

Fig. 4.19 Frequency response of grounded negative inductor for shunt/parallel RL load

It is obvious from Fig. 4.19 that the grounded negative inductor serves as an excellent neutralizing device for the parallel RL load impedance over the frequency range starting at 200.0 MHz and an upper limit of 800.0 MHz.

Next, **impmatchnegimpconv** is executed interactively to design a negative capacitor for a series RC load:

```
$ ./impmatchnegimpconv i
frequency band low limit (MHz)
75
frequency band high limit (MHz)
500
load reactance capacitor|inductor (pF|nH)
10
load resistance (Ohm)
50
load reactance type capacitor|inductor (1|2)
1
load configuration series|parallel RC/RL (1|2)
2
SPICE netlist text file : negimpcapind.cir
```

The corresponding frequency response is shown in Fig. 4.20.

Figure 4.20 shows that as signal frequency increases, the output signal amplitude decreases in a linear and monotonic way, very similar to that of the frequency response of a parallel RC (transformed equivalent series RC) Zobel network, as shown in Fig. 4.16.

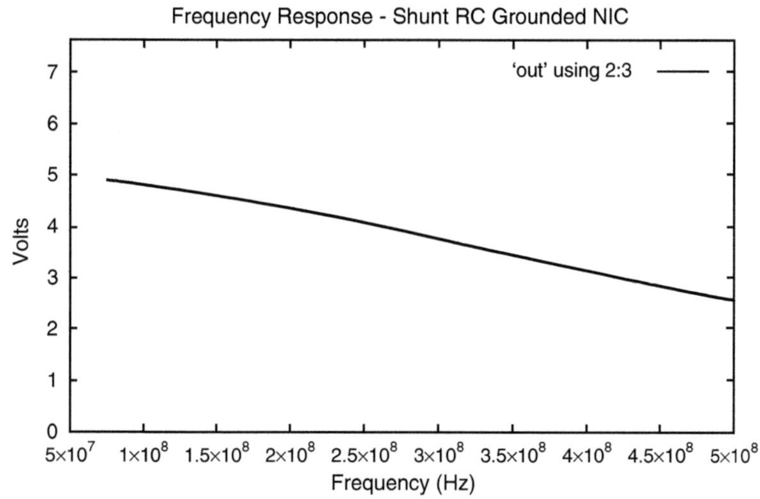

Fig. 4.20 Frequency response for negative capacitor for shunt RC load impedance

The performance of the grounded negative capacitor for the shunt RC load can be improved by exploring the design space, i.e., judiciously adjusting component value, e.g., using a different bi-junction NPN transistor, as shown in the following SPICE netlist snippet, which shows a higher bandwidth transistor replacing the original one in the negative capacitance:

```
*XQ0 3 1 4 BFR92A
*XQ1 1 3 5 BFR92A
Q0 3 1 4 BFS17
Q1 1 3 5 BFS17
```

Finally, **impmatchnegimpconv** is used interactively to design a floating negative inductor for a series RL load:

```
$ ./impmatchnegimpconv i
frequency band low limit(MHz)
100
frequency band high limit(MHz)
800
load reactance capacitor|inductor (pF|nH)
12.5
load resistance(Ohm)
50
load reactance type capacitor|inductor (1|2)
2
load configuration series|parallel RC/RL (1|2)
1
SPICE netlist text file : negimpcapind.cir
```

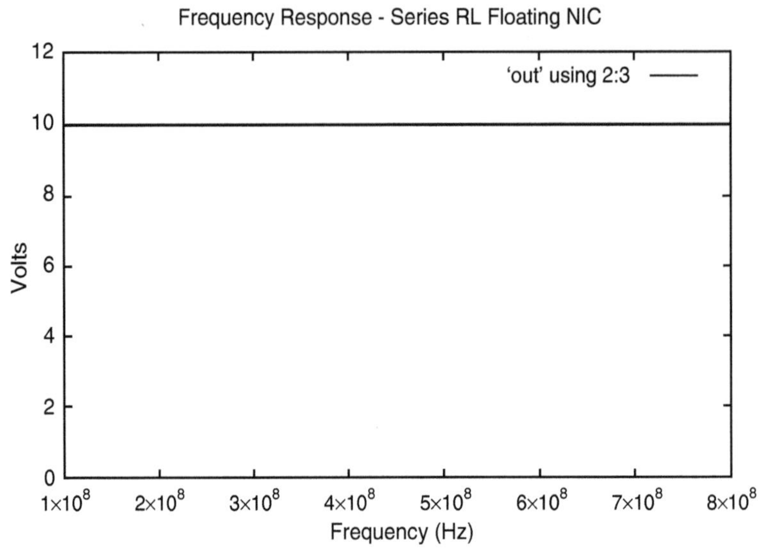

Fig. 4.21 Frequency response for floating negative inductor for floating series RL

The frequency response is shown in Fig. 4.21.

The floating negative inductor provides a near perfect impedance matching for the series RL load impedance.

4.7 Broad/Wide Band Impedance Matching with Unequal Source and Load Resistance Norton Single Element Series/Shunt Transforms – Automated Design and SPICE Performance Analysis

The analysis of broadband impedance matching schemes examined so far relied on the assumption that the load and source resistances are equal—this is a subset of real-world design cases, when the source and load resistances are unequal. Unequal source–load impedance matching is tackled by using the Norton subnetwork, which consists of a small network of capacitors and inductors, combined with an ideal transformer. The ideal transformer is the perfect broadband impedance transformer. As it is impossible to construct an ideal transformer for RF and microwave frequencies, it is either absorbed or replaced by another reactive network. In this subsection, the Norton series, shunt impedance transformers are analyzed/examined in detail.

There are two points to keep in mind while using the Norton subnetwork scheme:

- A Norton subnetwork may be inserted in any location in the network being analyzed.

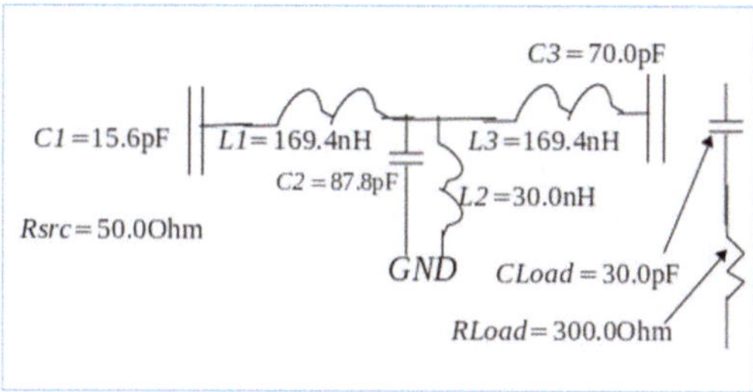

Fig. 4.22 Simple band-pass filter to match a 50-ohm source impedance and a series complex load consisting of series 30.0 pF capacitor and 300.0-ohm resistor

- A Norton subnetwork of any type (series, shunt, or more complicated ones) may be inserted as many times as the designer wishes to in the target network.

Consequently, the design process cannot be fully automated from start to end; only the steps that involve complicated numerical computations are automated. The Norton shunt transform is illustrated with a simple design example. Consider the simple band-pass filter in Fig. 4.22.

The shunt capacitor *C2* (87.8 pF) is selected for replacement by an equivalent Norton subnetwork. There is nothing special about this choice. The component values for the Norton subnetwork elements are calculated by a simple C computer language program **impmatchbbnsrsh,** which can be invoked in both the interactive and batch/command line argument modes. Typing **./impmatchbbnsrsh** at the command prompt generates the help message:

```
$ ./impmatchbbnsrsh
incorrect/insufficient arguments
interactive mode
./impmatchbbnsrsh i
batch/command line argument mode
./impmatchbbnsrsh b|B|c|C
<band center or operating frequency(MHz)>
<reactance to be replaced(pF/nH)>
<load resistance(Ohm)>
<source resistance(Ohm)>
<reactance type capacitor/inductor (1/2)>
<reactance configuration series/shunt(1/2)>
```

Using the value of 87.8 picoFarad for *C2*, the computed values for the corresponding T Norton subnetwork are:

```
$ ./impmatchbbnsrsh i
band center or operating frequency(MHz)
175
reactance (pF/nH)
87.8
load resistance(Ohm)
300
source resistance(Ohm)
```

```
50
reactance type capacitor/inductor (1/2)
1
reactance configuration series/shunt(1/2)
2
transformer turns ratio 1:0.408248
SPICE netlist template : IMMNortinA.cir
```

The contents of the SPICE template file are:

```
Norton series/shunt sub-network unequal source/load impedance

.PARAMS FLLIM=1.750000e+06 FHLIM=2.187500e+08
.PARAMS RSRC=50.000000 RLOAD =300.000000 AMPL=10

** NORTON SUBNETWORK TO BE ABSORBED
.SUBCKT 1 2 TNORTON
** 1 IN
** 2 OUT
C0 1 3 -1.272652e-10
C1 3 0 2.150652e-10
C2 3 2 3.117348e-10
.ENDS

** INSERT FINALIZED IMPEDANCE
** MATCHING HERE
** AC(SMALL SIGNAL) ANALYSIS
VSIG 1 0 DC 0.001 AC {AMPL}
RS 1 2 {RSRC}
RL 3 0 {RLOAD}
XIMP 2 3 IMPMATCHBP

.OPTIONS MRYJOD=GEAR NOPAGE RELTOL=1m MINBREAK=5ps
.AC LIN 50000 {FLLIM} {FHLIM}
.PRINT AC V(3)
.END
```

When *C2* in Fig. 4.22 is replaced by the corresponding T Norton subnetwork (check SPICE netlist above), the resulting intermediate circuit is shown in Fig. 4.23.

In the next step, the ideal transformer is moved to the far right, adjacent to and in series with the load resistor. The reactive elements that were to the right of the ideal transformer (Fig. 4.23) are scaled and/or absorbed. With reference to Fig. 4.23, the two capacitors *C1* and *C2* are combined to give a capacitor of value 13.88 pF. Again, with reference to Fig. 4.19, the inductor *L2* is scaled as $\frac{30.05}{0.408*0.408} = 144.98\ \ 1\text{nH}$ and the inductor *L4* as $\frac{169.4}{0.408*0.408} = 1017.637\text{nH}$. The series capacitor pair consisting of the load capacitor and the *C5* (with reference to Figs. 4.18 and 4.19) are combined and scaled as $0.408*0.408*\left(\frac{30.0*15.58}{30.0+15.59}\right) = 1.707$ pF. Finally, the load resistor becomes $\frac{50.0}{0.408*0.408} = 300.365$ ohm. Combining all these values, the final band-

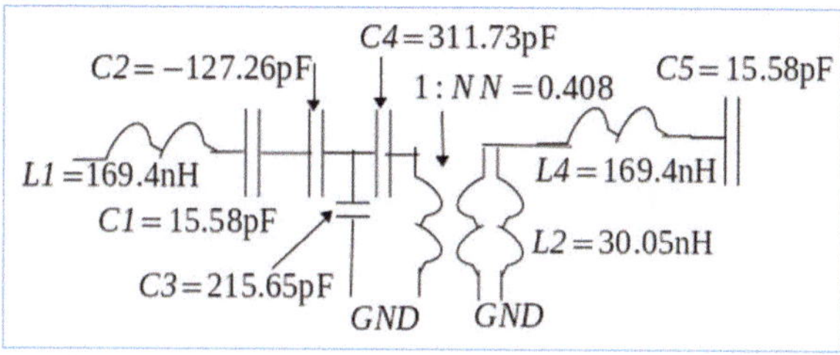

Fig. 4.23 Capacitor *C2* of Fig. 4.18 replaced by equivalent shunt Norton T network. Note that each of the Norton T subnetwork elements has been identified by having arrows pointing to them

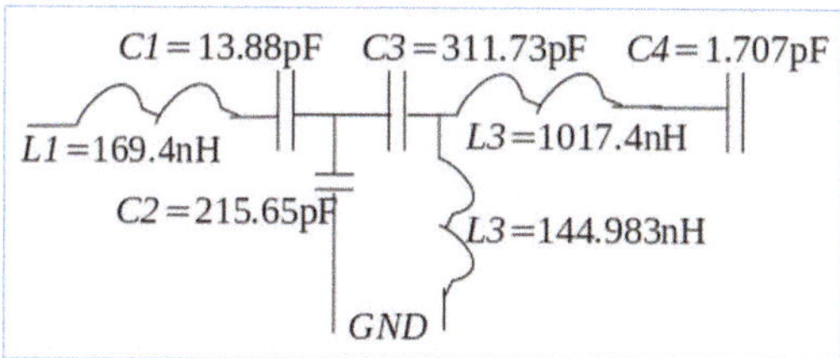

Fig. 4.24 Finalized band-pass filter with 40.0-ohm source resistance and 300.0-ohm load resistance

pass filter with a source resistor of 50.0 ohms and load resistor of 300.0 ohms is shown in Fig. 4.24.

Based on Fig. 4.26 above, the SPICE netlist for the Norton shunt transformed band-pass filter is:

```
.SUBCKT IMPMATCHBP 1 2
** 1 IN
** 2 OUT
C0 3 4 13.88E-12
C1 8 0 215.65E-12
C2 9 5 311.73E=12
C3 6 2 1.707E-12
L0 1 3 169.4E-9
L1 5 0 144.98E-9
L2 5 6 1017.4E-9
R0 4 8 10.0
R1 4 9 10.0
.ENDS
```

Note that the two extra resistors R0 and R1 have been inserted to force convergence of the calculations executed by the SPICE simulator. This netlist is inserted into the SPICE template AC or small signal analysis, which generates the frequency and phase responses, as shown in Figs. 4.25 and 4.26.

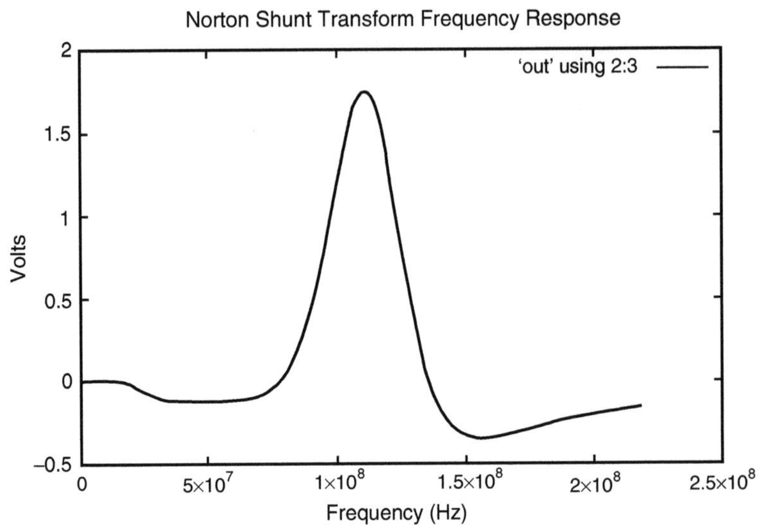

Fig. 4.25 Frequency response of Norton shunt transformed band-pass filter to match unequal source–load resistances

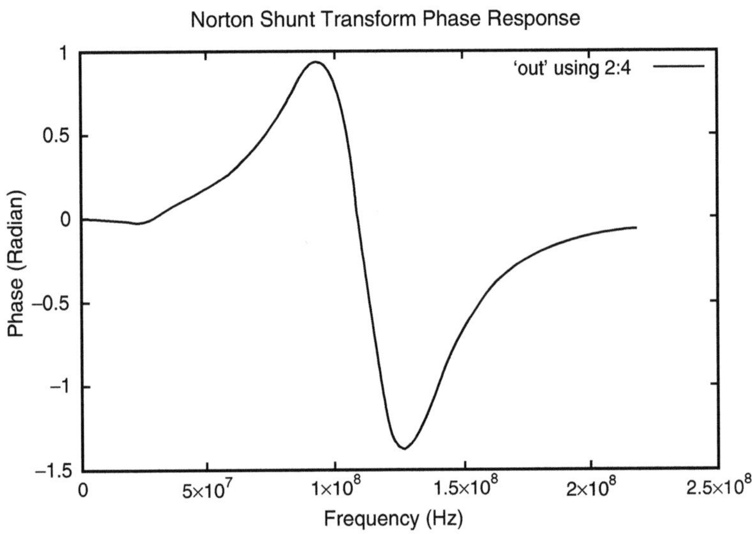

Fig. 4.26 Phase response of Norton shunt transformed band-pass filter to match unequal source–load resistors

4.8 Broad/Wide Band Impedance Matching with Unequal Source and Load Resistance General Norton to Element PI, T Transforms – Automated Design and SPICE Performance Analysis

A more general form of Norton transformations is examined in detail. The starting subnetwork that is transformed consists of an ideal transformer and a pair of series/shunt capacitors or inductors, not in mixed mode. *Keeping in mind that the Norton transforms may be applied as many times as the designer so wishes and at any appropriate location in the candidate network, only the detailed numerical calculation steps can be automated.* Consider the simple band-pass filter that is to be used to impedance match a 1.0-ohm source with a 50.0-ohm load (Fig. 4.27).

The first step is to add an ideal 1:*N* transformer to the left of the load resistor, as shown in Fig. 4.28. *N* is the square root of the load, source resistances and may be less than 1 but always positive.

To exploit the Norton transforms listed in Chap. 3, the ideal transformer is moved to the right of *L1*, as in Fig. 4.29. Then each of *C1*, *C2*, and *L2* is scaled by *N*, as shown.

The key step in the design process is to replace the ideal transformer and the two scaled capacitors $\frac{C1}{N^2}$ and $\frac{C2}{N^2}$ with a PI or T network consisting of only capacitors. This key design calculation is performed by the C computer language executable **impmatchnrtneqv,** which does the following calculations:

- Determines the values of the capacitors/inductors for the equivalent PI/T network that would replace the ideal transformer and the two scaled capacitors
- Formats most of the results as a text SPICE input netlist file, including a test bench for AC (small) analysis, as well as calculation of reflection coefficient, insertion loss, and standing wave ratio (SWR).

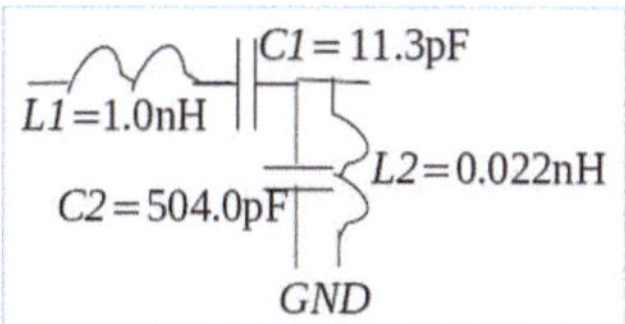

Fig. 4.27 Simple two-stage band-pass filter to match a 1.0-ohm source with a 50.0-ohm load

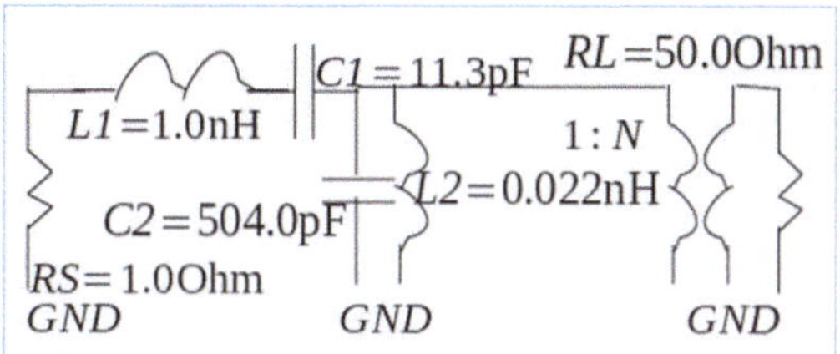

Fig. 4.28 Ideal transformer added to left of load resistance

Fig. 4.29 Ideal transformer shifted to new location

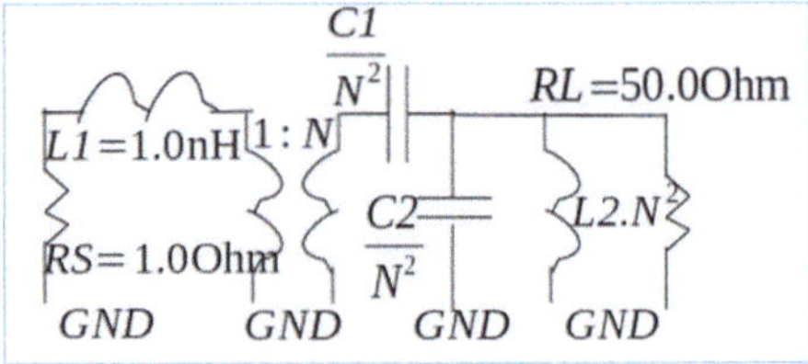

Impmatchnrtneqv may be executed interactively or in a batch/command line argument mode. Typing **./impmatchnrtneqv** generates help information. When executed interactively, it generates the following output:

```
./impmatchnrtneqv i pi 1
band center/cut-off/operating frequency(MHz)
500
load resistance(Ohm)
50
source resistance(Ohm)
1
series capacitor(pF)
11.3
shunt capacitor(pF)
504
turns ratio 1:0.141421
SPICE template netlist : IMMNeqv.cir
```

The contents of the SPICE template file are listed below. It is SPICE template netlist as

the finalized impedance matching subnetwork has to be added by hand before the AC analysis is done by the SPICE simulator:

```
Two initial element Norton transformers PI - T

.PARAMS FLLIM=5.000000e+06 FHLIM=6.250000e+08 AMPL=10
.PARAMS RS=1.000000 RL=50.000000

.SUBCKT NRTNSUBNET1 1 2
** 1 IN
** 2 OUT
Cs 1 2 1.130000e-11
Cs 2 0 5.040000e-10
.ENDS

.SUBCKT NRTNPIEQV1 1 2
** 1 IN
** 2 OUT
C0 1 0 -6.860307e-11
C1 2 0 2.568510e-08
```

```
C2 1 2 7.990307e-11
.ENDS

** ENTER FINALIZED IMPEDANCE TRANSFORMER COMPONENTS
.SUBCKT NRTNPIEQV1 1 2
** 1 IN
** 2 OUT
R0 5 0 5.0
R1 6 0 5.0
R2 7 0 5.0
.ENDS

** AC(SMALL SIGNAL) ANALYSIS
VSIG 1 0 DC 0.0001 AC {AMPL}
RS 1 2 {RS}
RL 3 0 {RL}
XM0 2 3 NRTNPIEQV1

.OPTIONS METHOD=GEAR NOPAGE RELTOL=1m MINBREAK=5ps
.AC LIN 50000 {FLLIM} {FHLIM}
*.PRINT AC V(3)
** REFLECTION COEFFICIENT ETC.,
.PRINT AC V(2)
.END
```

Note that negative capacitors or inductors do not exist in practice but can be tackled easily. A negative capacitor must have a positive double-valued capacitor connected in parallel, and a negative-valued inductor must have a positive double-valued inductor connected in series. Alternatively, the negative-valued capacitor/inductor can be replaced by its identical positive-valued counterpart.

The finalized SPICE input format Norton transformed impedance matching network, consisting of only capacitors and inductors, is:

```
.SUBCKT NRTNPIEQV1 1 2
** 1 IN
** 2 OUT
C0 3 5 6.860307e-11
C1 4 6 2.568510e-08
C2 3 4 7.990307e-11
L0 1 3 1.0e-9
L1 2 7 44.55e-9
R0 5 0 5.0
R1 6 0 5.0
R2 7 0 5.0
R3 4 2 1.0
.ENDS
```

The resistors have been added to force the SPICE simulation engine to converge quickly The completed circuit, with the load and source resistors is shown in Fig. 4.30,

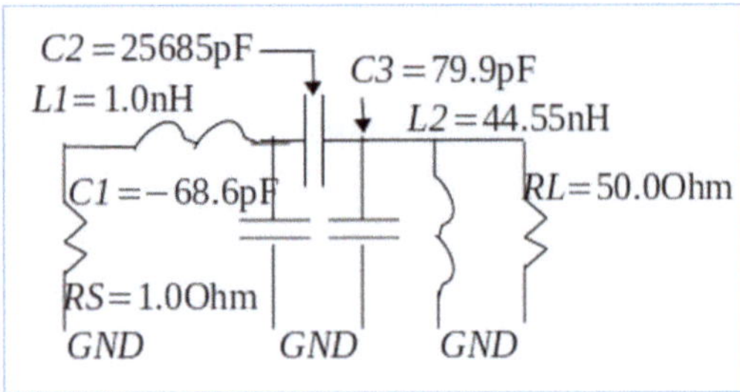

Fig. 4.30 Finalized band-pass impedance matching network with unequal load and source resistors and equivalent Norton subnetwork (components pointed to by arrows)

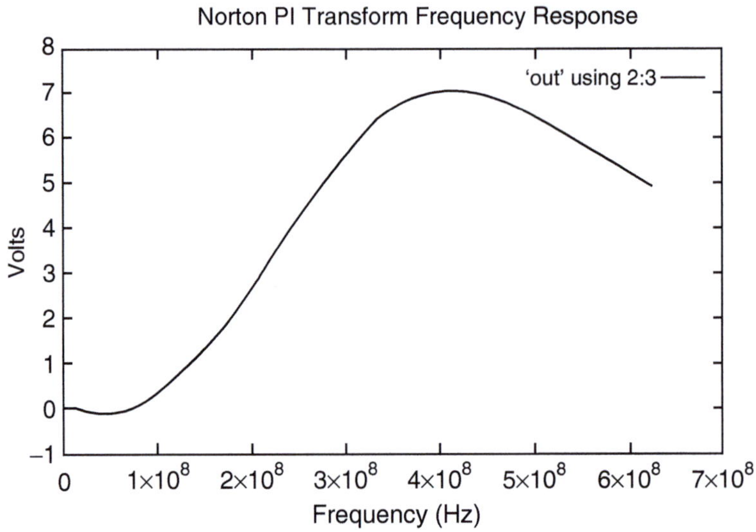

Fig. 4.31 Frequency response broadband unequal source–load resistance Norton transformed impedance matching network

The frequency and phase responses are shown in Figs. 4.31 and 4.32, respectively.

4.8.1 Key Observation

Owing to the Bode–Fano inequalities (in turn dependent on the frequency dependency of both capacitive and inductive reactances), the fundamental step to achieve source–load impedance matching is to neutralize capacitive and inductive reactances, i.e., *"convert" complex impedances to real ones.* This key step can be completed with the Zobel transform or by using negative capacitors or inductors, as examined in detail in Sects. 4.5 and 4.6 of this chapter:

- The simple 1 pole diplexer (Zobel subnetwork/transform) effectively "converts" any series RC or series RL impedance to a pure resistor. Parallel RC or RL

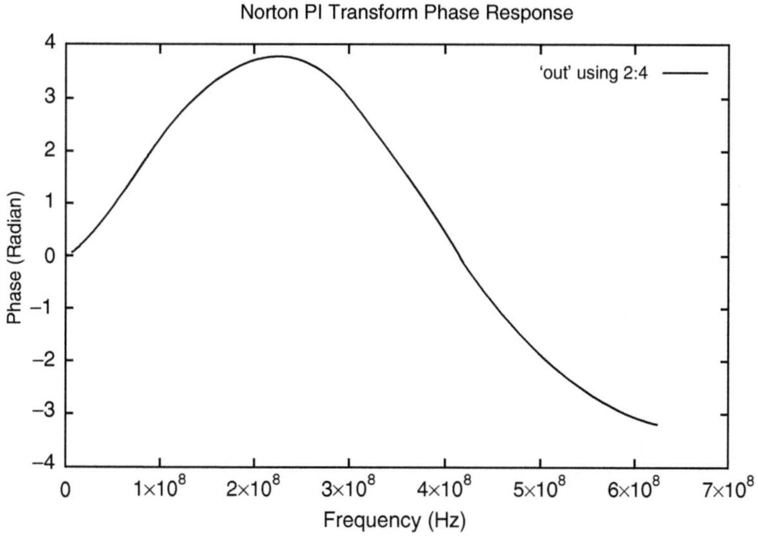

Fig. 4.32 Phase response broadband Norton PI impedance transformed impedance matching network with unequal source–load resistance

impedances can be transformed into their equivalent series RC or RL counterparts for analysis.

- An equally viable alternative to the Zobel transform is to use negative capacitors or inductors. This scheme *does not* require that a parallel RC or RL impedance be transformed into its equivalent series RC or RL form for analysis. The only minor drawback is that the designer must ensure that the operating frequency range of the bi-junction transistor or field-effect transistor (FET) (to be used to construct the negative capacitor or inductor) has an operating frequency range that matches the required operating frequency range over which the negative capacitor inductor is supposed to work.
- The Norton subnetwork transform enables impedance matching of unequal (complex or real) impedances, though it is best suited for real source and load impedances. So once the reactive element of a complex impedance has been neutralized ("converted" into a pure resistor using the Zobel subnetwork or negative impedance transform), any two unequal source–load resistor pairs can be impedance matched using the Norton subnetwork scheme.

Clearly, any complex or pure real impedance can be impedance matched using a combination of these three techniques.

4.9 Broad/Wide Band Low-Pass Matching Network – Equal Source–Load Resistance – Parallel Load Capacitor – One-End Reflections – Automated Design and SPICE Performance Analysis

The C computer language executable **impmatchbblpA** computes component values for an *n*th order low-pass configuration broad/wide band impedance matching network that satisfies the following conditions:

- Both the load and source resistances are equal.
- The load consists of the resistor and either a parallel capacitor or a series inductor.

impmatchbblpA can be invoked in either the interactive mode or batch/command line argument mode. Typing **./impmatchbblpA** at the command prompt generates the help information as:

```
$ ./impmatchbblpA
incorrect/insufficient arguments ...
interactive parallel capacitor-resistor load
./impmatchbblpA i prc
interactive series inductor-resistor load
./impmatchbblpA i slr
batch/command line argument mode ...
./impmatchbblpA b|B|c|C
<cut-off frequency(MHz)>
<load reactance(picoFarad/nanoHenry)>
<load resistance(Ohm)>
<parallel capacitor|series inductor-resistor(1/2)>
<filter order>
even/odd low pass filters of maximum
order 8/9 are supported
sample batch/command line input
./impmatchbblpA B 450 10 50 1 6
This scheme is for parallel capacitor and
series inductor-resistor loads only
```

Using the sample command line input generates the following output, with some intermediate results being shown first, the load consisting of the resistor and parallel capacitor:

```
$ ./impmatchbblpA B 450 10 50 1 6
normalized coefficients
0  1.000000e+00
1  5.173817e-01
2  1.413650e+00
3  1.931508e+00
4  1.932332e+00
5  1.415902e+00
6  5.204579e-01
```

```
denormalized coefficients
0  5.000000e+01
1  3.661583e-12
2  2.501151e-08
3  1.366955e-11
4  3.418846e-08
5  1.002054e-11
6  9.208385e-09
load reactance absorbed
0  5.000000e+01
1  3.661583e-12
2  2.501151e-08
3  1.366955e-11
4  3.418846e-08
5  1.002054e-11
6  9.208385e-09
SPICE netlist file : LPimpmatch.cir
```

The contents of the SPICE input netlist file are:

```
Low pass filter impedance matching
* Identical load source resistance
* and shunt load capacitor

.PARAMS FREQ=8.287546e+07 LLIM=4.143773e+06 HLIM=1.450321e+08
AMPL=10

.SUBCKT LPFILT 1 2
* 1 IN
* 2 OUT
C1 1 0 3.661583e-12
L2 1 3 2.501151e-08
C3 3 0 1.366955e-11
L4 3 4 3.418846e-08
C5 4 0 1.002054e-11
L6 4 2 9.208385e-09
.ENDS

* COMMENT OUT TRANSIENT ANALYSIS
* TO RUN AC ANALYSIS AND VICE-VERSA

CLOAD 5 0 1.000000e-11
CCOMP 5 0 2.000000e-11
RSRC 1 2 50.000000
RLOAD 5 0 50.000000
XLP 3 4 LPFILT
VTST0 2 3 DC 0.0 AC 0.0
VTST1 4 5 DC 0.0 AC 0.0

* FOR TRANSIENT ANALYSIS
VSIG 1 0 DC 0.005 SIN(0 {AMPL} {FREQ} 0 0 0)
```

```
* FOR AC(SMALL SIGNAL) ANALYSIS
VSIG 1 0 DC 0.005 AC {AMPL})

.OPTIONS METHOD=GEAR NOPAGE RELTOL=1m MINBREAK=5ps
.IC
.TRAN 75.0ns 500.0us 50.0ns UIC
* INPUT/SOURCE SIDE
.PRINT TRAN V(3) I(VTST0)

* LOAD/OUTPUT SIDE
.PRINT TRAN V(4) I(VTST1)
.AC LIN 20000 {LLIM} {HLIM}
.PRINT AC V(2)
** FOR REFLECTION COEFFICIENT ETC.,
.PRINT AC V(2)
.END
```

The utility computer executables **rmscalc** and **refcoeffcalc** are used to compute, respectively, the RMS power delivered to the load from the source and the reflection coefficient and insertion loss:

```
RMS input current : 0.09578572 Amperes
RMS input voltage : 2.425093 Volts
Total number of sample data values : 523396
RMS output current : 0.08292143 Amperes
RMS output voltage : 3.103942 Volts
Total number of sample data points L 523396
```

So almost all RMS input signal power from source is delivered to the load.

The reflection coefficient and insertion loss are shown in Figs. 4.33 and 4.34. As expected from the Bode–Fano criteria, the bandwidth is vastly reduced, to keep the gain high.

The results for two additional parallel RC load design cases are presented next. The first uses an even order low-pass matching network, while the second uses an odd order low-pass matching network. The target cut-off frequencies are 350 MHz and 500 MHz, respectively, for the even and odd order cases. The output generated from the even order matching network is:

```
$ ./impmatchbblpA c 350 12 50 1 8
normalized coefficients
0  1.000000e+00
1  3.899854e-01
2  1.110644e+00
3  1.662386e+00
4  1.961298e+00
5  1.961919e+00
6  1.664155e+00
7  1.113291e+00
8  3.931091e-01
```

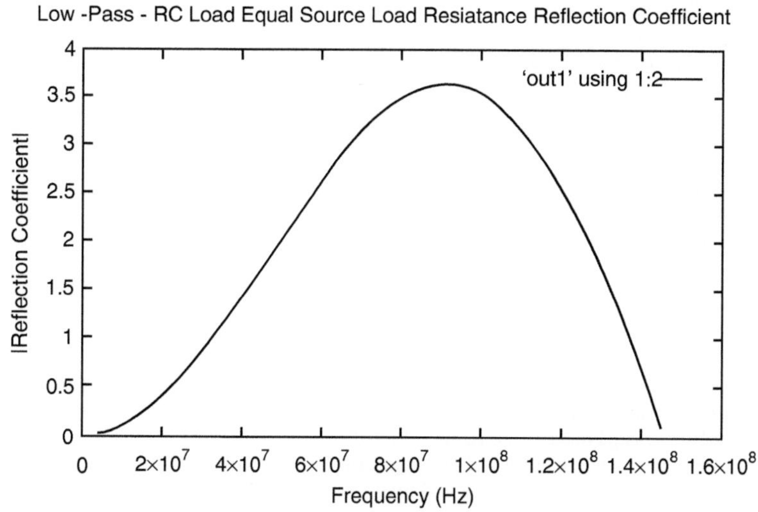

Fig. 4.33 Reflection coefficient low-pass impedance matching network

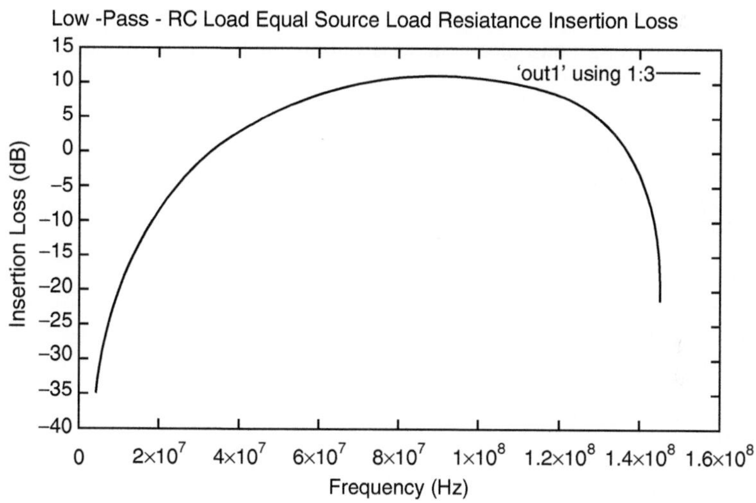

Fig. 4.34 Insertion loss low-pass impedance matching network

```
denormalized coefficients
0  5.000000e+01
1  3.548548e-12
2  2.526487e-08
3  1.512635e-11
4  4.461552e-08
5  1.785186e-11
6  3.785612e-08
```

```
7  1.013004e-11
8  8.942426e-09
load reactance absorbed
0  5.000000e+01
1  3.548548e-12
2  2.526487e-08
3  1.512635e-11
4  4.461552e-08
5  1.785186e-11
6  3.785612e-08
7  1.013004e-11
8  8.942426e-09
SPICE netlist file : LPimpmatch.cir

RMS input current : 0.09040181 Amperes
RMS input voltage : 2.686081 Volts
Total number sample data values : 657086
RMS output current : 0.07741149 Amperes
RMS output voltage : 2.686081 Volts
Total number of sample data values : 657088
```

The ratio of RMS power delivered to load to the input RMS power is 0.856304647.

The corresponding computer-generated output for the odd order matching network cases is:

```
$ ./impmatchbblpA c 500 10 75 1 7
normalized coefficients
0  1.000000e+00
1  4.448200e-01
2  1.246446e+00
3  1.801444e+00
4  1.999999e+00
5  1.802825e+00
6  1.248935e+00
7  4.479250e-01
denormalized coefficients
0  7.500000e+01
1  1.888832e-12
2  2.977180e-08
3  7.649443e-12
4  4.777069e-08
5  7.655309e-12
6  2.983126e-08
7  1.902017e-12
load reactance absorbed
0  7.500000e+01
1  1.888832e-12
2  2.977180e-08
3  7.649443e-12
4  4.777069e-08
5  7.655309e-12
```

```
6  2.983126e-08
7  1.902017e-12
SPICE netlist file : LPimpmatch.cir

RMS input current : 0.06351009 Amperes
RMS input voltage : 2.446184 Volts
Total number of sample data values : 570583
RMS output current : 0.05435744 Amperes
RMS output voltage : 2.446184 Volts
Total number of sample data values : 570584
```

The ratio of RMS power delivered to load to the input RMS power is 0.855886679.

For both cases, the ratio of output RMS power delivered to the load to the input RMS power can be increased by judiciously adjusting component values of the matching network, without touching the load or source impedances.

4.10 Low-Pass Configuration Matching Network Design with Bode–Fano Constraints – Automated Design and Evaluation of Continued Fraction Form of Driving Impedance Function

The C computer language executable **impmatchbbmnet** implements the Bode–Fano inequality (on gain bandwidth) based impedance matching network for complex loads, as discussed in Chap. 3. A drawback of this method is that synthesizing the final impedance matching network involves evaluation of the poles and zeros of ratios of high-order polynomials using the continued fraction technique, followed by frequency and impedance scaling of the generated prototype low-pass filter coefficients to create the finalized network. Evaluation of continued fraction expansion of ratios of high-order polynomials requires specialized mathematical techniques and software packages such as International Mathematical Standard Library (IMSL).

impmatchbbmnet can be executed in the command line argument mode or interactively. Typing **./impmatchbbmnet** at the command prompt generates the help information as:

```
$ ./impmatchbbmnet
incorrect/insufficient arguments
interactive mode
./impmatchbbmnet i
batch/command line argument mode
./impmatchbbmnet b|B|c|C
<frequency band low limit (MHz)>
<frequency band high limit (MHz)>
<load reactance 1 (uH)>
<load reactance 1 (pF)>
```

```
<load resistance(Ohm)>
<Resistance to be matched(Ohm)>
<pass band ripple(dB)>
<matching network order>
sample batch|command line input
./impmatchbbmnet b 75 125 1 12.5 200 50 0.1 4
```

Using the sample command line input from the command prompt generates the sample design as:

```
$ ./impmatchbbmnet b 75 125 1 12.5 200 50 0.1 4
Computed network parameters
Mean frequency 100.000000 MHz
Relative bandwidth 0.500000
Low frequency limit normalized 0.750000
High frequency limit normalized 1.250000
Mean normalized 1.030776
A parameter 0.500000
Load reactance 1 normalized 3.140000
Load reactance 2 normalized 1.570000
Load resistance normalized 1.000000
termination ratio 4.000000
Bode Fano integral values 2.000000   0.137937
epsilon 0.008519
maximum insertion loss calculated 0.092466

Poles and zeros
a 1.542207 b 0.519187
Primed poles and zeros
poles +/- -1.575313 + +/-j 1.730517
zeros +/- -0.383379 + +/-j 0.805530
Poles
-1.927759 + j -0.787656
-0.197241 + j 0.787656
Zeros
-1.465265 + j -0.191689
-0.659735 + j 0.191689
Reflection function
Denominator
(p^2 + -3.855517p + 0.620403) (p^2 + -0.394483p + 0.620403)
Numerator
(p^2 + -2.930530p + 0.036745) (p^2 + -1.319470p + 0.036745)
```

The fully multiplied out form of the numerator, of the reflection function, after converting to polar coordinates is $p^4 + 0.3927p^3 + 2.204p^2 + 0.4551p + 1.017$ and the corresponding denominator, after similar conversion to polar coordinates, is $p^4 + 1.668p^3 + 3.414p^2 + 2.765p + 27.5p + 1.585$. Then assuming that p is of the form $p = j \cdot w$, where w is the angular frequency, the numerator expression reduces to $1.017 + w^4 - 2.204w^2 + jw(0.4551 - 0.3927w^2)$, and the corresponding expression for the denominator is $1.585 + w^4 - 3.414w^2 + jw(2.765 - 1.668w^2)$. The absolute

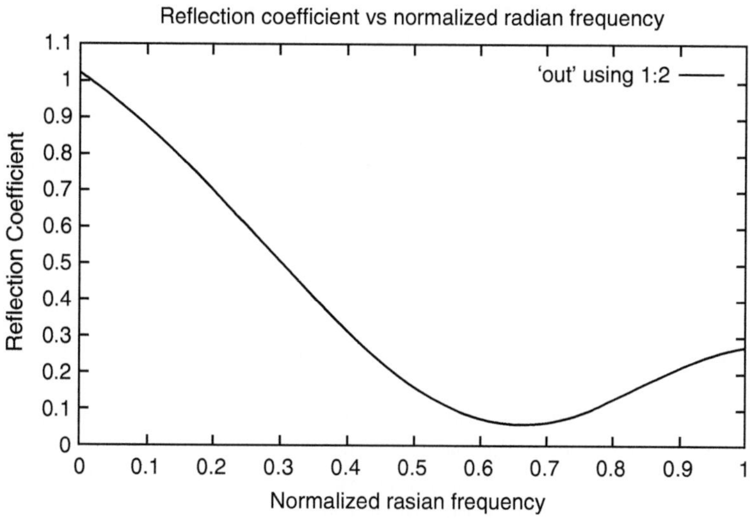

Fig. 4.35 Reflection coefficient versus normalized frequency

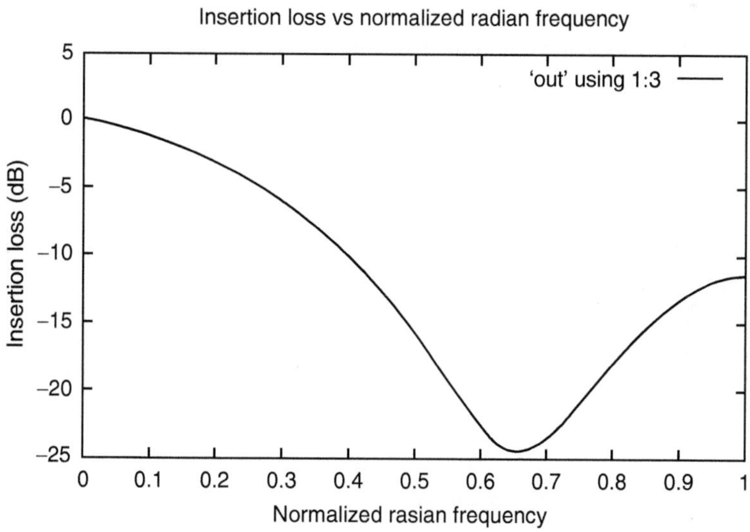

Fig. 4.36 Insertion loss versus normalized frequency

value of the reflection function, plotted as a function of normalized frequency, is shown in Fig. 4.35, while Fig. 4.36 shows the insertion loss versus the normalized frequency. Also, using the expressions examined in Chap. 3, the expression for the driving point impedance is obtained from the expression for the reflection function as:

$Z(p) = \frac{1 - \text{rho}}{1 + \text{rho}} = \frac{0.6377p^3 + 0.6555p^2 + 1.144p + 0.339}{p^4 + 1.03p^3 + 1.86p^2 + 1.51p + 1.356}$, where rho is the reflection function. To synthesize the physical impedance matching network, the expression for the driving impedance needs to be expanded as a continuous fraction. This is achieved with the Viscovatov algorithm. This algorithm, which is related to the Padé algorithm, states that the ratio of two high-order polynomials of the form $F(x) = \frac{P(x)}{Q(x)} = \frac{a_{10} + a_{11}x + a_{12}x^2 + a_{13}x^3 + a_{14}x^4 + ..}{a_{00} + a_{01}x + a_{02}x^2 + a_{03}x^3 + a_{04}x^4 + ..}$ can be expanded as a continuous fraction whose terms are of the form (Rogers) $\frac{a_{10}}{a_{00}} \frac{a_{20}x}{a_{10}} \frac{a_{30}x}{a_{20}}$ and so on, where the general expression for $a_{m,n}$ is: $a_{m,n} = a_{m-1,0} \cdot a_{m-2,n+1} - a_{m-2,0} \cdot a_{m-1,n+1}$.

The C computer language executable **impmatchcontfracexp** implements the Viscovatov algorithm and returns the continuous fraction terms in the Rogers format. It can be executed in both the interactive and batch/command line input modes. When executed in the batch/command line input mode, it reads in the polynomial coefficients from a simple tab-separated variable text file. In this text input file, the coefficients for the 0th power of the variable are placed in the very first line, followed by the coefficients of the next higher power and so on. At present, **impmatchcontfracexp** only supports a maximum of 5th order polynomials in the denominator and numerator. The first column of the input text file contains the coefficients for the denominator and the second column the coefficients for the numerator, both formatted starting with the coefficient of the 0th power of the variable, followed by the next higher power and so on, as mentioned above. The input file (named polycoeffs) to **impmatchcontfracexp** for the driving point impedance function shown above is:

```
1.356   0.339
1.61   1.155
2.86   0.6555
1.03    0.6377
1.0   0.0
```

Typing **./impmatchcontfracexp** at the command prompt generates the help information as:

```
$ ./impmatchcontfracexp
incorrect/insufficient arguments
interactive mode
./impmatchcontfracexp i
batch/command line argument mode
./impmatchcontfracexp b|B|c|C polycoeffs
'polycoeffs' is name of tab separated variable input file in text format
column 1 - denominator polynomial coefficients
column 2 - numerator polynomial coefficients
sample input file - tab separated variable
204   25
```

```
245   8
55    1
15    0
1     0
```

Now using the input file polycoeffs listed earlier, the continued fraction expansion of the driving point impedance function is:

```
$ ./impmatchcontfracexp B polycoeffs
0.250000
0.254000
25.613339
-1.432378
0.505249
-0.657549
3.628693
0.741205
```

Finally, the continued fraction expansion yields the normalized component values for the impedance matching network, which need to be frequency and impedance scaled to obtain the component values for the capacitor, inductor, and resistor values for the impedance matching network. This is left as an exercise for the reader. **impmatchcontfracexp** is continuously being improved to handle higher order polynomials, and fine-tuned versions will be available in the near future.

4.11 Non-optimized Real Frequency Method – Automated Design and Numerical Integration Evaluation of Hilbert Transform

The C computer language executable **impmatchbbrefros** implements the nonoptimized real frequency impedance matching technique, as discussed in Chap. 3, Sects. 3.10 and 3.11. One of the key steps in this calculation is the use of numerical integration to evaluate the Hilbert transform of a set of intermediate data. Only one optimization step is used at the very last step because ill-conditioned matrices cause most optimization algorithms to not converge. Also, as the name “real frequency” suggests, a complete set of values of real and imaginary parts of the load impedance must be generated, or provided, over the entire operating frequency range, often difficult to obtain.

impmatchbbrefros can be invoked interactively or via the command line in the batch mode. Typing **./impmatchbbrefros** at the command prompt generates the help message as:

```
$ ./impmatchbbrefros
incorrect/insufficient arguments
```

```
interactive mode ...
./impmatchbbrefros i
batch or command line argument mode...
./mpmatchbbrefros b|B|c|C
<frequency band low limit (MHz) <
<frequency band upper limit (MHz) >
<load reactance capacitor (pF) |inductor (nH) >
<load resistance (Ohm) >
<transducer power gain flatness (e.g., 0.75) >
<load type parallel capacitor|inductor (1|2) >
<Chebyschev polynomial degree>
sample command line input
./impmatchbbrefros b 150 500 10 50 0.7851 1 6
```

Using the sample input from the command prompt generates a list of values of the real and imaginary parts of the driving point impedance as follows:

```
$ ./impmatchbbrefros B 150 500 10 50 0.7851 1 6
0  0.433893  2.196863
1  0.443579  1.514558
2  0.454344  1.335819
3  0.466211  1.397218
4  0.479203  1.429308
5  0.493346  1.461780
6  0.508661  1.494607
7  0.525174  1.527759
8  0.542906  1.561206
9  0.561880  1.594912
10  0.582116  1.628840
11  0.603631  1.662947
12  0.626443  1.697191
13  0.650564  1.731523
14  0.676005  1.765894
15  0.702774  1.800249
16  0.730874  1.834531
17  0.760304  1.868681
18  0.791060  1.902634
19  0.823133  1.936325
20  0.856506  1.969684
21  0.891162  2.002638
22  0.927072  2.035114
23  0.964206  2.067033
24  1.002526  2.098315
25  1.041985  2.128880
26  1.082532  2.158643
27  1.124108  2.187519
28  1.166645  2.215422
29  1.210068  2.242265
30  1.254295  2.267959
31  1.299234  2.292418
32  1.344786  2.315554
..........................................
```

```
.........................................
89  0.086562  4.365579
90  0.046683  4.690487
91  0.021706  5.050764
92  0.026784  5.446812
93  0.059313  5.871454
.........................................
.........................................
101  1.662949  10.547387
102  2.110870  11.300138
103  2.634204  12.088222
104  3.240267  12.909612
105  3.936874  13.761516
106  4.732364  14.640240
107  5.635627  15.541002
108  6.656132  16.457730
.........................................
.........................................
119  28.998395  23.240871
120  32.417451  22.652451
121  36.144096  21.514423
122  40.199443  19.630408
123  44.605742  16.727329
124  49.386426  12.549767
125  54.566165  1.516583
126  60.170916  -13.341662
127  66.227974  -46.450915
```

Here, the output data set contains 128 data values, with the first column indicating the index, the second the minimum resistance (details in Chap. 3, Sect. 3.11), and the last the reactance value, obtained after Hilbert transform of the resistance.

4.12 Guanella 1:*n* Ideal and Lossy Transmission Line Transformer Impedance Matching – Real Load Automated Design and SPICE Performance Analysis

The C computer language program **impmatchguanella** does all the design calculations and generates the text SPICE input netlist and test bench (small signal and transient analysis) for both ideal and lossy Guanella transmission line transformer for impedance matching of real load. Typing **./impmatchguanella** at the command prompt generates the help information:

```
$ ./impmatchguanella
incorrect/insufficient arguments
interactive mode ...
1:n voltage ratio transmission line xfrmr
./impmatchguanella i 1:n
n:1 voltage ratio transmission line xfrmr
./impmatchguanella i n:1
```

```
batch/command line argument mode
1:n voltage ratio transmission line xfrmr
./impmatchguanella b|c|B|C 1:n
<frequency band lower limit(GHz)>
<frequency band upper limit(GHz)>
<load impedance(Ohm)>
<source impedance(Ohm)>
<relative permittivity - inter conductor insulation>
sample command line input for 1:n Guanetta transmission line
transformer
./impmatchguanella B 1:n 0.25 0.75 90 10 1
identical argument list for n:1 voltage ratio
transmission line transformer
```

Using the sample input from the command line generates:

```
$ ./impmatchguanella B 1:n 0.25 0.75 90 10 1
1:1 balun Z0 30.000000 Ohm
1:1 balun physical length 0.086603m
SPICE netlist test file balungu.cir
```

The contents of the generated SPICE netlist are:

```
GUANELLA 1:3 TRANSMISSION LINE TRANSFORMER

.PARAMS FREQ=4.330127e+08 Z0=30.000000
.PARAMS LLIM=2.165064e+07 HLIM=6.495191e+09 AMPL=10 TD=2.886751e-10
.PARAMS R=0.0175 L=15.0E-9 G=0.0
.PARAMS C=86.0E-12 LEN=0.086603

.MODEL LTXL LTRA R={R} L={L} G={G} C={C} LEN={LEN}

.SUBCKT GUANELLA1_3 1 2 3 4
** 1 IN 1
** 2 IN 2
** 3 OUT 1
** 4 OUT 2
T0 1 2 3 5 Z0={Z0}
T1 1 2 5 6 Z0={Z0}
T2 1 2 6 4 Z0={Z0}
.ENDS

.SUBCKT GUANELLALOSSY1_3 1 2 3 4
** 1 IN 1
** 2 IN 2
** 3 OUT 1
** 4 OUT 2
O0 1 2 3 6 LTXL
O1 1 2 6 6 LTXL
O2 1 2 6 4 LTXL
.ENDS
```

```
VSIG 1 0 DC 0.0002 SIN(0 {AMPL} {FREQ} 0 0 0)
RS 1 2 10.000000
RL 4 5 90.000000
VTST0 2 3 DC 0.0 AC 0.0
VTST1 5 6 DC 0.0 AC 0.0
XGU 3 0 4 6 GUANELLALOSSY1_3

* COMMENT OUT TRANSIENT ANALYSIS SECTION
* TO RUN AC ANALYSIS AND VICE VERSA
** TRANSIENT ANALYSIS

.OPTIONS METHOD=GEAR NOPAGE RELTOL=1m MINBREAK=5ps

.IC V(3)=0.25 V(4)=0.3
.TRAN 100.0ps 10.0us 50.0ps UIC
* INPUT/SOURCE SIDE
.PRINT TRAN V(3) I(VTST0)
* OUTPUT/LOAD SIDE
.PRINT TRAN V(4) I(VTST1)

** AC(SMALL SIGNAL ANAYSIS
** FOR REFLECTION COEFFICIENT, RETURN LOSS and VSWR
VSIG 1 0 DC 0.001 AC {AMPL}
.OPTIONS METHOD=GEAR NOPAGE RELTOL=1m MINBREAK=5ps
.AC LIN 20000 {LLIM} {HLIM}
.PRINT AC V(2)
.END
```

Using the AC or small signal analysis in the above SPICE netlist, the plot of the reflection coefficient with the *ideal* transmission line segments in the Guanella 1:3 impedance transformer is shown in Fig. 4.37.

When the 1:3 Guanella transmission line transformer is constructed with the lossy transmission line (whose capacitance/length, inductance/length, etc.) are listed in the SPICE netlist, the reflection coefficient and insertion loss are shown in Fig. 4.38 and 4.39, respectively.

Now the value for the inductance/length for the lossy transmission line is modified as:

```
NEW : .PARAMS L=10.0E-10 G=0.0 : OLD : .PARAMS L=15.0E-9 G=0.0
```

The new reflection coefficient is shown in Fig. 4.40 and the corresponding insertion loss in Fig. 4.41.

This is a clear demonstration of how sensitive the performance characteristics of the transmission line transformer are to the key transmission line parameters.

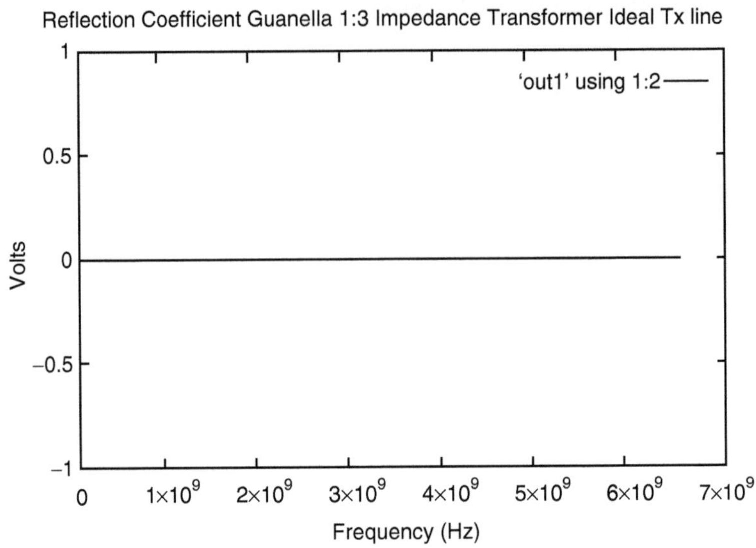

Fig. 4.37 Reflection coefficient – ideal transmission line 1:3 Guanella TXLT

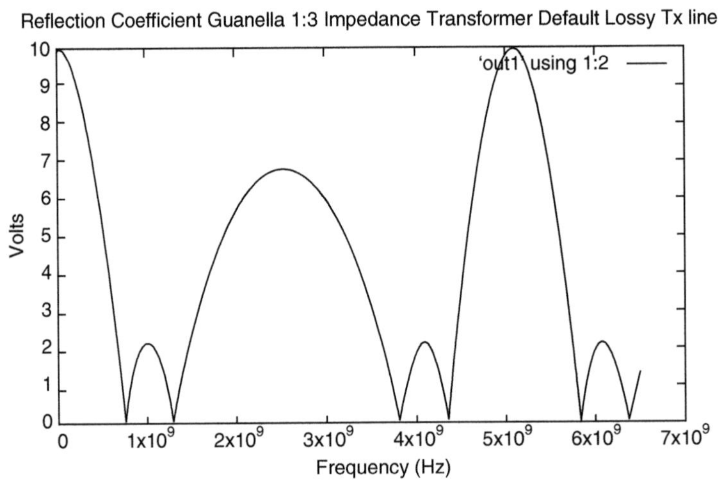

Fig. 4.38 Reflection coefficient – Guanella 1:3 lossy transmission line transformer

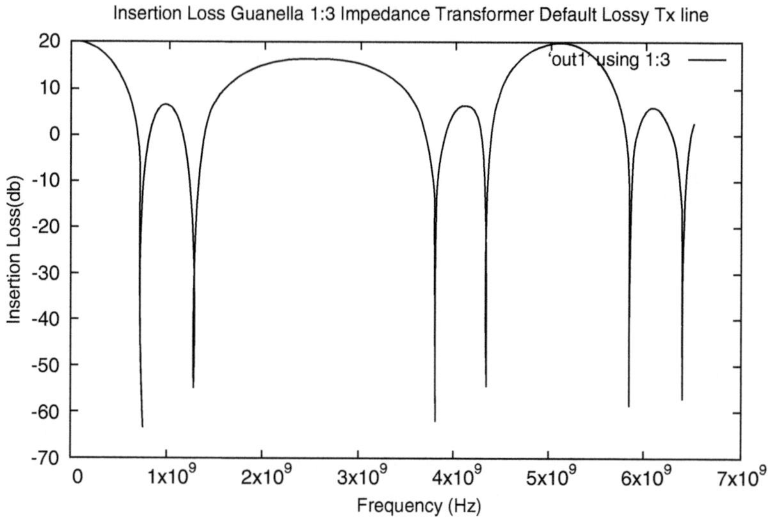

Fig. 4.39 Insertion loss – lossy transmission line Guanella 1:2 TXLT

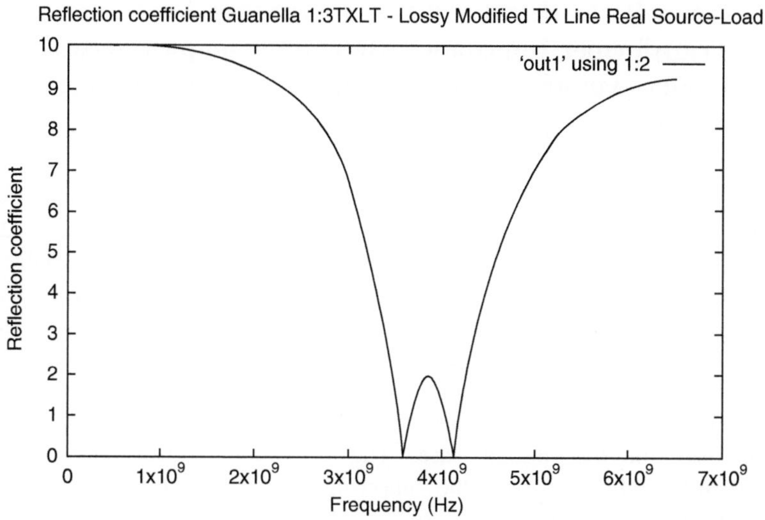

Fig. 4.40 Reflection coefficient – modified lossy transmission line Guanella 1:2 TXLT

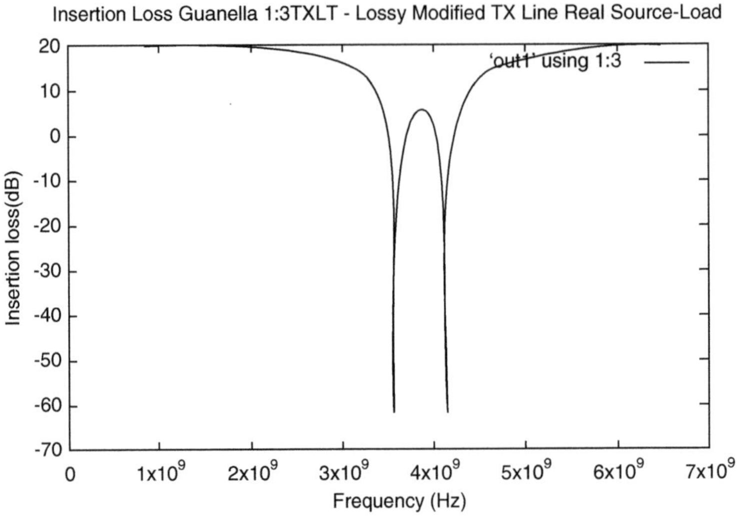

Fig. 4.41 Insertion loss – modified lossy transmission line Guanella 1:2 TXLT

4.13 Guanella 1:*n* Ideal and Lossy Transmission Line Transformer Impedance Matching Complex Series Load – Automated Design and SPICE Performance Analysis

Most real-world devices have complex input/output impedances (parallel/series RC/RL). The Guanella transmission line transformer as originally conceived and designed was for real source–load impedances. A simple work-around enables the Guanella transmission line transformer to tackle complex loads and involves using the equivalent Zobel subnetwork to "convert" the complex load into a simple resistor. The 90.0-ohm resistor load used in the previous design example (Sect. 4.12) load is converted into a complex load with a series 10.0 pF capacitor, and the equivalent Zobel subnetwork is constructed. This equivalent "resistor" is then used as the load for this design case. The C computer language executable **impmatchbbpszobel** generates the following SPICE netlist for the equivalent Zobel subnetwork for the series RC load:

```
$ ./impmatchbbpszobel b 350 750 10 90 1 1
series RC | RL load
SPICE netlist file : serparzobel.cir
```

The contents of the SPICE netlist file are:

```
Series RC - RL Zobel network

.PARAMS FLLIM=3.500000e+08 FHLIM=7.500000e+08 AMPL=10

.SUBCKT SERZOBEL 1 2
** 1 IN
** 2 OUT
R0 2 3 90.000000
C0 3 0 1.000000e-11
R1 1 2 1
R2 1 4 90.000000
LCOMP 4 0 1.234568e-15
.ENDS

VSIG 1 0 DC 0.001 AC {AMPL}
RS 1 2 90.000000
XZ0 2 3 SERZOBEL

.OPTIONS METHOD=GEAR NOPAGE RELTOL=1m MINBREAK=5ps
.AC LIN 50000 {FLLIM} {FHLIM}
.PRINT AC V(3)
.END
```

The frequency response of this equivalent Zobel subnetwork, corresponding to the series 90.0 ohm and 10.0 pF capacitor, is shown in Fig. 4.42.

Now the SPICE netlist for a Guanella 1:3 transmission line impedance transformer is generated using the C computer language executable **impmatchguanella** with a source load of 10.0 ohms and a load impedance of 90.0 ohms. *Then the real*

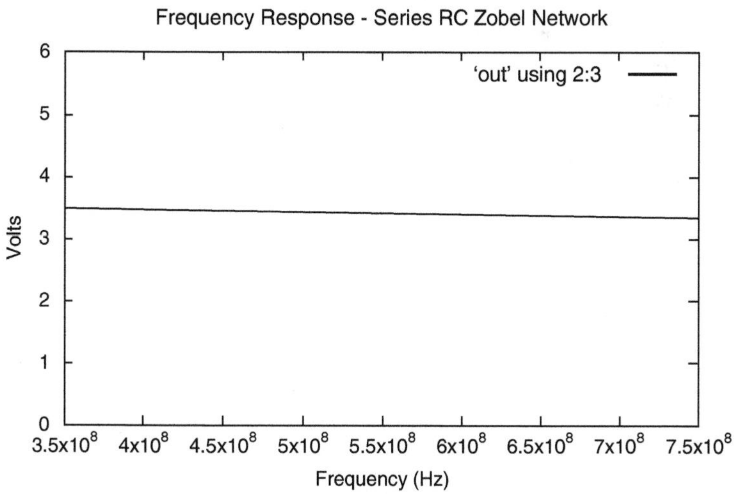

Fig. 4.42 Frequency response of Zobel subnetwork corresponding to series RC

load of the Guanella impedance transformer is replaced by the Zobel network for the series RC impedance as:

```
GUANELLA 1:3 TRANSMISSION LINE TRANSFORMER

.PARAMS FREQ=4.330127e+08 Z0=30.000000
.PARAMS LLIM=2.165064e+07 HLIM=6.495191e+09 AMPL=10 TD=2.886751e-10
.PARAMS R=0.0175 L=15.0E-9 G=0.0
.PARAMS C=86.0E-12 LEN=0.086603

.MODEL LTXL LTRA R={R} L={L} G={G} C={C} LEN={LEN}

.SUBCKT GUANELLA1_3 1 2 3 4
** 1 IN 1
** 2 IN 2
** 3 OUT 1
** 4 OUT 2
T0 1 2 3 5 Z0={Z0}
T1 1 2 5 6 Z0={Z0}
T2 1 2 6 4 Z0={Z0}
.ENDS

.SUBCKT GUANELLALOSSY1_3 1 2 3 4
** 1 IN 1
** 2 IN 2
** 3 OUT 1
** 4 OUT 2
O0 1 2 3 6 LTXL
O1 1 2 6 6 LTXL
O2 1 2 6 4 LTXL
.ENDS
.
.SUBCKT SERZOBEL 1 2
** 1 IN
** 2 OUT
R0 2 3 90.000000
C0 3 0 1.000000e-11
R1 1 2 1
R2 1 4 90.000000
LCOMP 4 0 1.234568e-15
.ENDS

RS 1 2 10.000000
*RL 4 5 90.000000
** REPLACE REAL LOAD WITH ZOBEL NETWORK
XRL 4 5 SERZOBEL
VTST0 2 3 DC 0.0 AC 0.0
VTST1 5 6 DC 0.0 AC 0.0
XGU 3 0 4 6 GUANELLA1_3
```

The results for the reflection coefficient and insertion loss with ideal transmission line Guanella 1:3 impedance matching subnetwork driving the Zobel transformed series RC load are shown in Figs. 4.43 and 4.44.

Clearly, the reflection coefficient vs. *frequency curve of* Fig. 4.43 *is identical to the reflection coefficient* vs. *frequency plot of a quarter wave transmission line impedance transformer used to impedance match a real source with a real load.*

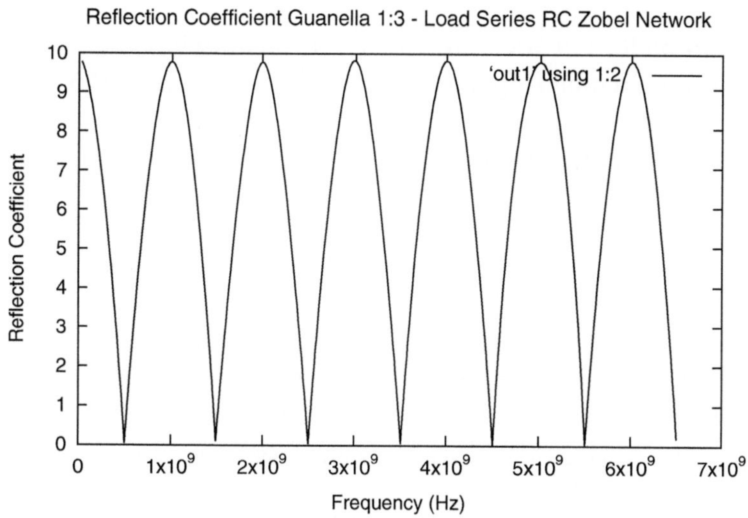

Fig. 4.43 Frequency response of Zobel subnetwork corresponding to series RC load

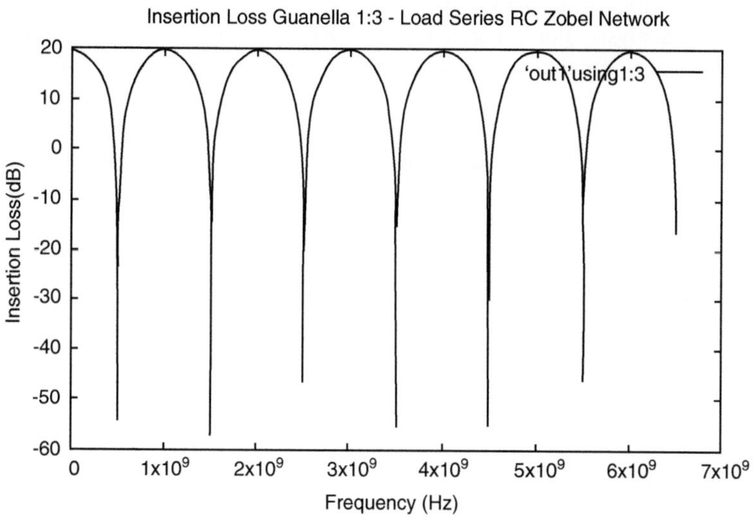

Fig. 4.44 Insertion loss Zobel subnetwork corresponding to series RC load

This demonstrates that the Zobel equivalent subnetwork has successfully "converted" a complex series RC load into a resistor.

Identical analysis can be done with an n:1 Guanella impedance matching transformer with a series RL load. In addition, the analysis can be made more realistic by using a lossy transmission line mode, instead of the ideal model. These are left as exercises for the reader.

4.14 Guanella 1:n Ideal and Lossy Transmission Line Transformer Impedance Matching – Complex Parallel RL Load – Automated Design and SPICE Performance Analysis

Continuing with the current discussion thread, the problem of impedance matching a complex impedance consisting of a parallel inductor and resistor is now examined in detail. The two key initial design steps are:

- Transform parallel RL load to equivalent series RL load.
- Construct the Zobel subnetwork that would transform the equivalent series RL load (corresponding to original parallel RL load) into a pure resistor.

Finally, the Guanella 1:n impedance transformer drives the equivalent Zobel subnetwork.

The C computer language executable **impmatchbbpszobel** accepts as input the values of the resistor and capacitor or inductor for a parallel or series RC or RL pair. For a series RC or RL, it generates the component values for the corresponding Zobel network, formats the results as a SPICE netlist, and adds a test bench for AC or small signal analysis. For a parallel RC or RL impedance, it first computes the component values for the equivalent series RC or RL and then uses these freshly generated values for constructing the Zobel network and test bench, finally formatting all the results as a SPICE netlist.

Executing **impmatchbbpszobel** in the batch/command line argument mode with the following input for a parallel load consisting of a 160-ohm resistor and a 15.0-nanoHenry inductor generates:

```
$ ./impmatchbbpszobel b 250 750 15 160 2 2
parallel RC | RL load
SPICE netlist file : serparzobel.cir
```

The contents of the SPICE netlist are:

```
PARALLEL RC - RL Zobel network

.PARAMS FLLIM=2.500000e+08 FHLIM=7.500000e+08 AMPL=10
```

```
.SUBCKT PARZOBEL 1 2
** 1 IN
** 2 OUT
R0 2 3 9.764197
L0 3 0 1.408461e-08
R1 1 2 1
R2 1 4 9.764197
CCOMP 4 0 1.477310e-10
.ENDS

VSIG 1 0 DC 0.001 AC {AMPL}
RS 1 2 9.764197
XZ0 2 3 PARZOBEL

.OPTIONS METHOD=GEAR NOPAGE RELTOL=1m MINBREAK=5ps
.AC LIN 50000 {FLLIM} {FHLIM}
.PRINT AC V(3)
.END
```

The frequency responses of the Zobel network is shown in Fig. 4.45.

To design the Guanella 1:4 transmission line impedance transformer, the C computer language executable **impmatchguanella** is run from the command line with the following input parameter set:

Note that the selected frequency band is identical to that used in the AC analysis of the equivalent series RL load above. The generated SPICE input netlist for the Guanella 1:4 impedance transformer is:

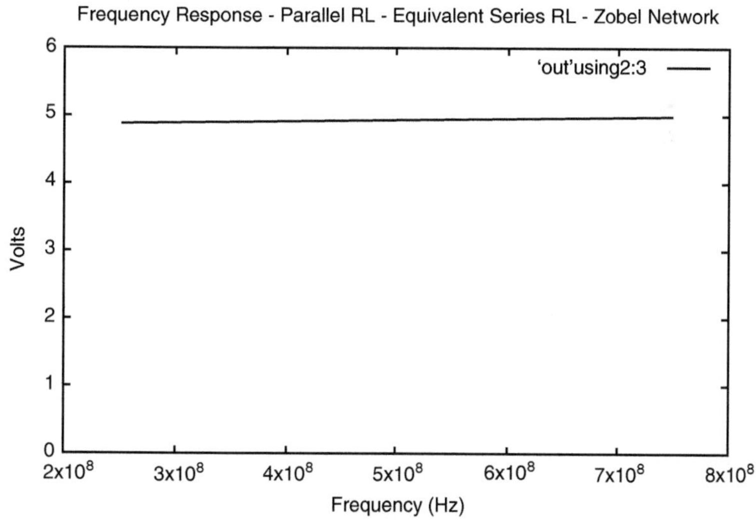

Fig. 4.45 Frequency response of Zobel subnetwork corresponding to series RL impedance transformed from original parallel RL impedance

```
$ ./impmatchguanella B 1:n 0.25 0.75 160 10 1
1:1 balun Z0 40.000000 Ohm
1:1 balun physical length 0.086603m
SPICE netlist test file balungu.cir
```

The contents of the generated SPICE netlist balungu.cir are:

```
GUANELLA 1:4 TRANSMISSION LINE TRANSFORMER

.PARAMS FREQ=4.330127e+08 Z0=40.000000
.PARAMS LLIM=2.165064e+07 HLIM=6.495191e+09 AMPL=10 TD=2.886751e-10

.PARAMS R=0.0175 L=15.0E-9 G=0.0
.PARAMS C=86.0E-12 LEN=0.086603
.MODEL LTXL LTRA R={R} L={L} G={G} C={C} LEN={LEN}

.SUBCKT GUANELLA1_4 1 2 3 4
** 1 IN 1
** 2 IN 2
** 3 OUT 1
** 4 OUT 2
T0 1 2 3 5 Z0={Z0}
T1 1 2 5 6 Z0={Z0}
T2 1 2 6 7 Z0={Z0}
T3 1 2 7 4 Z0={Z0}
.ENDS

.SUBCKT GUANELLALOSSY1_4 1 2 3 4
** 1 IN 1
** 2 IN 2
** 3 OUT 1
** 4 OUT 2
O0 1 2 3 7 LTXL
O1 1 2 7 7 LTXL
O2 1 2 8 8 LTXL
O3 1 2 8 4 LTXL
.ENDS

VSIG 1 0 DC 0.0002 SIN(0 {AMPL} {FREQ} 0 0 0)
RS 1 2 10.000000
RL 4 5 160.000000
VTST0 2 3 DC 0.0 AC 0.0
VTST1 5 6 DC 0.0 AC 0.0
XGU 3 0 4 6 GUANELLA1_4

** COMMENT OUT TRANSIENT ANALYSIS SECTION
** TO RUN AC ANALYSIS AND VICE VERSA
** TRANSIENT ANALYSIS
.OPTIONS METHOD=GEAR NOPAGE RELTOL=1m MINBREAK=5ps
```

```
.IC V(3)=0.25 V(4)=0.3
.TRAN 100.0ps 10.0us 50.0ps UIC
* INPUT/SOURCE SIDE
.PRINT TRAN V(3) I(VTST0)
* OUtPUT/LOAD SIDE
.PRINT TRAN V(4) I(VTST1)

** AC(SMALL SIGNAL ANAYSIS
** FOR REFLECTION COEFFICIENT, RETURN LOSS and VSWR
VSIG 1 0 DC 0.001 AC {AMPL}
.OPTIONS METHOD=GEAR NOPAGE RELTOL=1m MINBREAK=5ps
.AC LIN 20000 {LLIM} {HLIM}
.PRINT AC V(2)
.END
```

To analyze the performance characteristics of the Guanella 1:4 impedance matching circuit with the complex parallel RL load (160.0 ohms parallel with 15.0nH inductor), the Guanella SPICE netlist is modified by adding the following subcircuit, that is, a parallel combination of the original parallel RL load and the series R and C compensating pair from the Zobel network above. This is a key step. This hybrid load is used in place of the real 160.0-ohm load of the Guanella impedance matching circuit.

```
.SUBCKT PARZOBELHYBRID 1 2
** 1 IN
** 2 OUT
R0 2 0 160.0
L0 2 0 15e-08
R1 1 2 1
R2 1 4 9.764197
CCOMP 4 0 1.477310e-10
.ENDS
```

The modified Guanella 1:4 transmission line impedance transformer is now executed with the SPICE simulator to extract the curves for the reflection coefficient and insertion loss (Figs. 4.46 and 4.47).

It is obvious that the reflection coefficient vs. *frequency curve of* Fig. 4.47 *is identical to the reflection coefficient* vs. *frequency plot of a quarter wave transmission line impedance transformer used to impedance match a real source with a real load. This demonstrates that the hybrid Zobel equivalent subnetwork has successfully "converted" a complex parallel RL load to a resistor. It is essential in case of parallel RC or RL load impedances to use the hybrid Zobel circuit since the designer cannot change the parallel RC or RL configuration of the load impedance, and so any viable method to neutralize the reactive component of the given parallel RC or RL must include it as part of the solution. This is not necessary for the series RC or RL.*

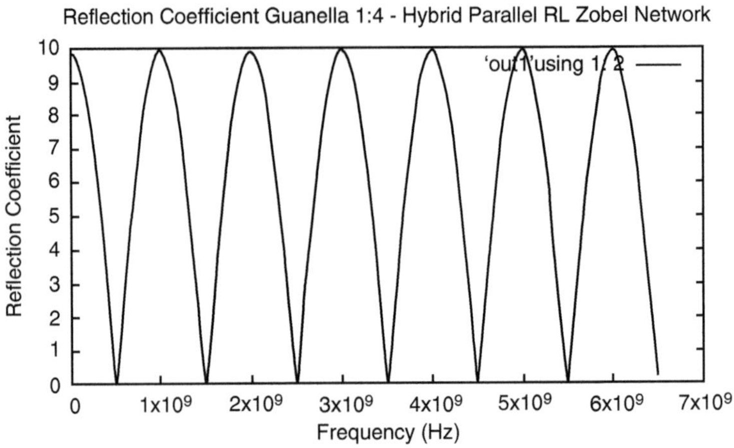

Fig. 4.46 Reflection coefficient Guanella 1:4 impedance transformer with hybrid Zobel network load

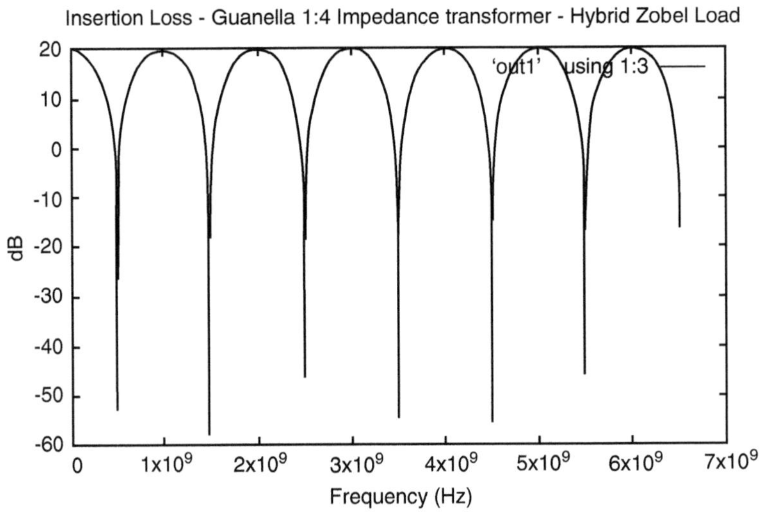

Fig. 4.47 Insertion loss Guanella 1:4 impedance transformer with hybrid Zobel network load

4.15 Simple Broad/Wide Band Impedance Matching Scheme

This scheme is simple and guaranteed to produce meaningful, usable results:

- For a series RC or RL impedance, straightforward application of Zobel subnetwork transform "converts" it to a resistor. A parallel RC or RL impedance must first be transformed into its equivalent series RC and RL counterpart for analysis purposes only. How to apply the Zobel subnetwork scheme to real-world physical circuit has been discussed/examined in detail in Sect. 4.6 of this chapter.

- Alternatively, a negative capacitor or inductor can be used to neutralize the capacitor or inductor (series or parallel RC or RL), "converting" a complex impedance into a real-valued resistor. The designer must ensure that the operating frequency range (of the bi-junction transistor or FET to be used to implement the negative capacitor or inductor) matches the predefined operating frequency range of the negative capacitor or inductor.
- Any two members of any unequal source–load resistor pair (resulting from "converting" or "neutralizing" a complex impedance) can be impedance matched using the Norton subnetwork scheme.

Exercises

- A series RC or RL impedance may be "converted" into a pure resistance by connecting a short-circuited transmission line segment of appropriate length in parallel. Derive expressions for the characteristic impedance and electrical and physical lengths for such a transmission line segment.
- The single transmission line segment narrow band impedance transformer has been simulated with an ideal transmission line model. Repeat the design with a nonideal transmission line. Explore design space by adjusting the capacitance/length and inductance/length of the nonideal transmission line to match its results with those of the ideal transmission line.
- The real frequency technique, as discussed in this chapter, does not use any optimization algorithm. Explore why this is the case. In addition, write a C computer language code to implement the conjugate gradient scheme.
- The SPICE simulations for impedance matching complex loads with the Guanella transmission liner transformer used the ideal transmission line model. Repeat the same simulations with the provided lossy transmission line mode. Explore design space by adjusting lossy transmission line parameters (capacitance/length, inductance/length) to force the reflection coefficient curve to look like the reflection coefficient curve obtained from the ideal transmission line model.
- Designing impedance matching networks starting from the Bode–Fano inequalities typically involves continued fraction expansion of ratios of high-order polynomials as the key last step. Download and install the IMSL libraries to perform this analysis for the design example of Sect. 4.10 of this chapter.
- Construct a 1:4.368 Guanella transmission line transformer.
- Explain why the frequency response plot of the Zobel network for a series RC or equivalent series RC has a very small slope, unlike the case for a series RL impedance. The same slope is noticed in active impedance matching of similar RC or equivalent series RC impedances.

Appendix A: Supplied C Computer Language Executables for the Linux and Windows Operating Systems

The C computer language executable modules come in two sets—one for the Linux operating system and the other for the Windows operating system. Using the executable modules for the Linux operating system is very simple and straightforward. However, to use the executable modules for the Windows operating system, the popular Linux emulator for Windows, MingW (*M*inimalist *GNU* for *Windows*) must first be downloaded and installed on the computer running the Windows operating system. Details for installing the MingW package are discussed in Appendix B.

impmatchdirl calculates component (capacitor, inductor) values for and generates the SPICE input netlist for the narrow band "L" impedance matching subcircuit for complex source-load impedance pairs.

impmatchtxline calculates the characteristic impedance and electrical and physical lengths of a single transmission line segment to impedance match a pair of complex source-load impedances, along with the SPICE netlist and test bench.

impmatchtxlstub does the design calculations to design and evaluate the performance characteristics (SPICE AC or small signal analysis) of single/double closed/open circuit parallel/series transmission line stub impedance matching subcircuit using both ideal and lossy transmission line models.

impmatchbbpszobel computes the component values for the Zobel network for a given parallel or series complex impedance(RC or RL), as well as the text SPICE netlist for AC or small signal analysis.

impmatchnegimpconv uses designer-supplied SPICE RF and microwave bi-junction transistor models to generate the corresponding text SPICE netlists for both the floating and grounded negative impedance converters for a given capacitor or inductor.

impmatchbbnsrsh calculates the component values for the Norton subnetwork elements and formats the results as a SPICE subcircuit for efficient performance analysis.

impmatchnrtneqv calculates the values of the capacitors/inductors for the equivalent PI/T network that would replace the ideal transformer and reactive element

A. Banerjee, *Automated Broad and Narrow Band Impedance Matching for RF and Microwave Circuits*, https://doi.org/10.1007/978-3-319-99001-9

pair. It formats the results as a text SPICE input netlist file, including a test bench for AC (small small) analysis, as well as calculation of reflection coefficient, insertion loss, and standing wave ratio (SWR).

impmatchbblpA computes component values for a low-pass configuration broad/wide band impedance matching network for which both the load and source resistances are equal and the load consists of a resistor and either a parallel capacitor or a series inductor.

impmatchbbmnet implements the Bode-Fano inequality (on gain bandwidth) based impedance matching network for complex loads. An auxiliary C computer language executable ***impmatchcontfracexp*** uses the Viscovatov algorithm to perform the continued fraction expansion of ratios of high-order polynomials.

impmatchbbrefros implements the nonoptimized real frequency impedance matching technique, including Hilbert transform but *not* the conjugate gradient method.

impmatchguanella does all the design calculations and generates the text SPICE input netlist and test bench (small signal and transient analysis) for both ideal and lossy Guanella transmission line transformers, for impedance matching of real load. Complex loads (parallel–series RC/RL) are first "converted" into real loads using ***impmatchbbpszobel.***

rmscalc computes the RMS (root-mean-squared) value of a set of variables. The input is a text file, consisting of maximum three columns of numbers. The numbers in each row consist of tab-separated real numbers. There is no limit on the number of rows.

refcoeffcalc computes reflection coefficient, insertion loss, and SWR (standing wave ratio) from a set of frequency response (amplitude, phase pairs) values. The input is a text file, as well as an output file. The user has to supply the amplitude of the signal used in the SPICE AC (small signal) analysis.

Each of these C computer programs is continuously being improved upon, and more efficient executables that generate more accurate results will be available in the near future.

Appendix B: How to Install MingW (GCC Compiler Suite) on Computers Running Windows Operating System

The popular, extremely reliable, and versatile GCC compiler suite can be easily installed on any Windows operating system computer, including the latest Windows 10. The following websites have detailed step-by-step instructions, with appropriate screenshots for installing GCC on computers running the Windows operating system, including the latest Windows 10.

http://www.codebind.com/cprogramming/install-mingw-windows-10-gcc/

http://www.mingw.org/wiki/Install_MinGW

http://www.mingw.org/wiki/howto_install_the_mingw_gcc_compiler_suite

https://www3.ntu.edu.sg/home/ehchua/programming/howto/Cygwin_HowTo.html

In addition, a number of YouTube videos also show the exact sequence of steps for achieving the same goal.

A. Banerjee, *Automated Broad and Narrow Band Impedance Matching for RF and Microwave Circuits*, https://doi.org/10.1007/978-3-319-99001-9

Index

B

Bode-Fano inequalities, 1, 5, 18, 19, 24, 26, 27, 60, 74, 101
Broadband impedance matching
  Bode-Fano conditions on gain-bandwidth, 17–19
  complex series RC/RL and parallel RC/RL load, 19–21
  deficiencies of conventional transformer, 32–33
  Hilbert transforms, 30–32
  ladder configuration low pass, 24–26
  non-zero reactance, 17, 26–27
  reactances/susceptances, 21–24
  RF and microwave FETs, 29–30
  transmission line, 34–35
  unequal source-load impedances, 27–29

C

Canonical impedances, 18
Canonical Norton transform sub-circuits, 29
C computer language
  impmatchbblpA, 76
  impmatchbbmnet, 81
  impmatchbbnsrsh, 67
  impmatchbbpszobel, 52, 56, 58, 92, 96
  impmatchbbrefros, 85
  impmatchcontfracexp, 84, 85
  impmatchdirl, 37, 38, 41
  impmatchguanella, 87, 97
  impmatchnegimpconv, 60, 62, 64, 65
  impmatchnrtneqv, 71
  impmatchpsrcrl, 57
  impmatchtxline, 43
  impmatchtxlstub, 47, 48
  L IMPEDANCE XFRMR, 39
  Limpmatchdir.cir, 41
  rmscalc and **refcoeffcalc**, 78
Conjugate gradient method, 30
Conjugate matching
  band impedance matching, 1
  broad and narrow band impedance, 5
  signal source with impedance, 4
  voltage wave, 4

D

Devices transmission gain, 23
Double series-shunt closed-open transmission line impedance, 14–16
Driving impedance, 81–85

F

Field effect transistor (FET), 29
Foster network, 32
Frequency dependency, 21, 28, 51, 74

G

Gain-bandwidth, 17–19, 21, 25, 26
Guanella and Ruthroff transmission line impedance transformer, 34–35

H

Hilbert transform, 30–32, 85–87

A. Banerjee, *Automated Broad and Narrow Band Impedance Matching for RF and Microwave Circuits*, https://doi.org/10.1007/978-3-319-99001-9

I
Ideal transformer
capacitors, 28, 29
ideal impedance matching device, 32–33
inductors, 28
left of load resiatance, 71
PI/T sub-network, 27, 28
in practice, 28
right of, 68
transformation ratio, 23
Ideal transmission line, 50, 51
impedance transformer, 46
parallel single closed circuit stub, 49
parallel single open circuit stub, 48
reflection coefficient, 90
single closed circuit stub
insertion loss, 51
reflection coefficients, 51
single open circuit stub
insertion loss, 50
reflection coefficients, 50
Impedance matching
Bode-Fano inequalities, 1
broad and narrow, 1
capacitor and inductor, 3
conductors, 2
conjugate matching, 4, 5
current/voltage wave, 3
signal power, 1
signal source, 2, 4
transmission line segment, 2, 4
types, 1
wave velocity, 2
International Mathematical Standard Library (IMSL), 26

K
Kirchoff's current law (KCL), 20

L
Lambda/12 transmission line impedance transformer, 11–12
L impedance transformer, 7–9
Load impedance
parallel/series RC/RL, 3
transmission line segment and source, 2
Lossy transmission line, 52, 53
insertion loss, 91
parallel single closed circuit stub, 49
parallel single open circuit stub, 49
reflection coefficient, 90
single closed circuit stub
insertion loss, 53
reflection coefficient, 53
single open circuit stub
insertion loss, 52
reflection coefficient, 52

M
Maximum power transfer, 1
Modified lossy transmission line
insertion loss, 92
reflection coefficient, 91

N
Narrow band impedance matching, 9–10
common L impedance transformer, 7–9
double series-shunt closed-open, 14–16
lambda/12 transmission line impedance transformer, 11–12
reflection coefficients, 10–11
single series shunt closed-open, 13–14
single transmission line segment (*see* Single transmission line segment)
source and load, 7
Negative impedance converter, 23, 24, 58, 60–62
Norton sub-network
canonical, 28
capacitors, 29
finalized bandpass impedance matching network, 74
PI and T sub-circuit forms, 28
series and shunt, 27
SPICE input, 73
sub-circuits, 28, 29
unequal source load impedance matching, 66
versatile techniques, 27

R
Real frequency impedance transform, 30–32
Reflection coefficient, 3, 4, 10, 18, 19, 25, 26, 43, 45, 46, 48, 71, 78, 89, 95, 99, 101
Root-mean-squared (RMS) power, 43, 45, 78, 80, 81

S
Single bifilar coil, 34
Single series shunt closed-open transmission line impedance, 13–14

Single transmission line segment
  characteristic impedance, 10
  conjugate matching condition, 10
  electrical length, 10
  signal source, 9
Source impedances, 7–9
SPICE netlist, 48, 49
  AC analysis section, 40
  batch/command line, 54, 56
  Bi-junction transistors in feedback, 60–66
  broad/wide band impedance matching, 71–75, 100–101
  broad/wide band low-pass matching network, 76–81
  contents of, 38
  design calculations, 37
  Fano constraints-automated design, 81–85
  floating series RL, 66
  grounded negative capacitor, 65
  grounded negative inductor, 63, 64
  Guanella transmission line transformer, 92
  ideal transmission line (*see* Ideal transmission line)
  ideal transmission line impedance transformer, 47
  ./impmatchbbnsrsh, 67
  impmatchbbpszobel, 53, 56, 58, 92, 96
  impmatchguanella, 87
  impmatchnegimpconv, 61, 62, 64, 65
  impmatchpsrcrl, 57
  inductor load reactance, 50–60
  L impedance matching sub-circuit, 37, 38
  Limpmatchdir.cir, 41
  lossy transmission line (*see* Lossy transmission line)
  lumped element circuits, 42
  negative impedance converter, 58, 62
  non-optimized real frequency method, 85–87
  Norton shunt transformed bandpass filter, 68–70
  Norton sub-network scheme, 66, 67, 75
  parallel RC/RL impedances, 74
  parallel/series capacitor, 50–60
  RC Zobel network, 58, 59
  RMS, 40, 48
  sample natch/command line, 48
  single/double closed/open circuit transmission line, 47–49
  single transmission line, impedance transformer, 43–47
  T Norton sub-network, 68
  transmission line impedance transformer, 93, 94, 97–100
  unmodified lossy transmission line segment, 45, 47
  Zobel network, 54, 55, 57
  Zobel sub-network, 93, 95, 97, 99
Standing wave ratio (SWR), 5, 43, 45, 48, 71

T

Telecommunication, 1
Test bench, 37, 38, 43, 47, 57, 58, 71, 87, 96
Transmission line
  characteristic impedances, 2, 3
  load impedance, 4
  physical length, 4
  unit length, 2

W

Wireless communication, 1

Z

Zobel network, 17, 19–21, 27, 35, 52, 54–59, 64, 94, 96, 97, 99–101
Zobel transform, 74, 95

MIX
Papier aus verantwortungsvollen Quellen
Paper from responsible sources
FSC® C105338

If you have any concerns about our products,
you can contact us on
ProductSafety@springernature.com

In case Publisher is established outside the EU,
the EU authorized representative is:
Springer Nature Customer Service Center GmbH
Europaplatz 3, 69115 Heidelberg, Germany

Printed by Libri Plureos GmbH
in Hamburg, Germany